chemical COMPOUNDS

NEIL SCHLAGER, JAYNE WEISBLATT, AND
DAVID E. NEWTON, *EDITORS*

Charles B. Montney, *Project Editor*

VOLUME 3

POLYSTYRENE-
ZINC OXIDE

U·X·L

*An imprint of Thomson Gale,
a part of The Thomson Corporation*

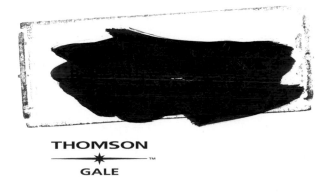

THOMSON
GALE

Detroit • New York • San Francisco • San Diego • New Haven, Conn. • Waterville, Maine • London • Munich

Chemical Compounds

Neil Schlager, Jayne Weisblatt, and David E. Newton, Editors

Project Editor
Charles B. Montney

Editorial
Luann Brennan, Kathleen J. Edgar, Jennifer Greve, Madeline S. Harris, Melissa Sue Hill, Debra M. Kirby, Kristine Krapp, Elizabeth Manar, Kim McGrath, Paul Lewon, Heather Price, Lemma Shomali

Indexing Services
Barbara Koch

Imaging and Multimedia
Randy Bassett, Michael Logusz

Product Design
Kate Scheible

Composition
Evi Seoud, Mary Beth Trimper

Manufacturing
Wendy Blurton, Dorothy Maki

LIBRARY OF CONGRESS CATALOGING-IN-PUBLICATION DATA

Weisblatt, Jayne.
    Chemical compounds / Jayne Weisblatt ; Charles B. Montney, project editor.
        v. cm.
    Includes bibliographical references and indexes.
    Contents: v. 1. Acetaminophen through Dimethyl ketone -- v. 2. Ethyl acetate through Polypropylene -- v. 3. Polysiloxane through Zinc oxide.
        ISBN 1-4144-0150-7 (set : alk. paper) -- ISBN 1-4144-0451-4 (v. 1 : alk. paper) -- ISBN 1-4144-0452-2 (v. 2 : alk. paper) -- ISBN 1-4144-0453-0 (v. 3 : alk. paper)
    1. Chemicals. 2. Organic compounds. 3. Inorganic compounds. I. Montney, Charles B., 1962- II. Title.
    QD471.W45 2006
    540--dc22   2005023636

This title is also available as an e-book
ISBN 1-4144-0467-0
Contact your Thomson Gale sales representative for ordering information.

Printed in China
10 9 8 7 6 5 4 3 2 1

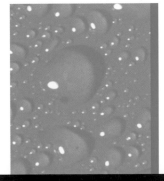

contents

Contents

Contents

Contents

volume 3

Contents

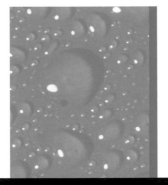

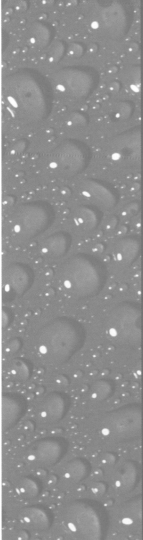

reader's guide

Water; sugar; nylon; vitamin C. These substances are all very different from each other. But they all share one property in common: They are all chemical compounds. A chemical compound consists of two or more chemical elements, joined to each other by a force known as a chemical bond.

This book describes 180 chemical compounds, some familiar to almost everyone, and some less commonly known. Each description includes some basic chemical and physical information about the compound, such as its chemical formula, other names by which the compound is known, and the molecular weight, melting point, freezing point, and solubility of the compound. Here are some things to know about each of these properties:

Other Names: Many chemical compounds have more than one name. Compounds that have been known for many centuries often have common names that may still be used in industry, the arts, or some other field. For example, muriatic acid is a very old name for the compound now called hydrochloric acid. The name remains in common use today. Marine acid and spirit of salt are other ancient names for hydrochloric acid, but they are seldom used in the modern world. All compounds have systematic names, names based on a set of rules devised by the International Union of Pure and Applied Chemistry (IUPAC). For example, the systematic name for the poisonous gas whose common name is mustard gas is 2,2'-dichlorodiethyl sulfide. When chemists talk about chemical compounds, they usually use only the official IUPAC name for a compound since that name leaves no doubt as to the substance about which they are talking. In some cases, a compound may have more than one official name, depending on the set of rules used in the naming process. For example, 1,1'-thiobis[2-chloroethane] is also an acceptable name for mustard gas. The "Other Names" section of each entry lists both the systematic (IUPAC) and common names for a compound.

Many compounds also have another kind of name, a brand name or trade name given to them by their manufacturers. For example, some trade names for the pain killer acetaminophen are Panadol™, Tylenol™, Aceta™, Genapap™, Tempra™, and Depacin™. The symbol next to each name means that the name is registered to the company that makes the compound. Trades names may be mentioned in the Overview or Uses sections of the entry for each compound.

Chemical Formula: A chemical formula is a set of symbols that tells the elements present in a compound and the relative numbers of each element. For example, the chemical formula for the compound carbon dioxide is CO_2. That formula tells that for every one carbon atom (C) in carbon dioxide there are two atoms of oxygen (O).

Chemists use different kinds of formulas to describe a compound. The simplest formula is a molecular formula. A molecular formula like CO_2 tells the kind and relative number of elements present in the compound. Another kind of formula is a structural formula. A structural formula provides one additional piece of information: The arrangement of elements in a compound. The structural formula for methanol (wood alcohol), for example, is CH_3OH. That formula shows that methanol consists of a carbon atom (C) to which are attached three hydrogen (H) atoms (CH_3). The carbon atom is also joined to an oxygen atom (O) which, in turn, is attached to a hydrogen atom (H).

Structural formulas can be written in a variety of ways. Another way to draw the structural formula for methanol, for example, is to show where individual bonds between atoms branch off other atoms in different directions. These structural formulas can be seen on the first page of nearly all entries in *Chemical Compounds*. In a third type of structural formula, the ball-and-stick formula, each element is represented by a ball of some size, shape, and/or color. The chemical bond that holds them together is represented by sticks. This can be represented on paper in a drawing that simulates a three-dimensional model, by computer software, or actually in three dimensions from a kit with balls and sticks.

All three kinds of structural formulas are given for each compound described in this book. The only exception is some very large compounds known as polymers that contain many hundreds or thousands of atoms. In such cases, the formulas given shown only one small segment of the compound.

Compound Type: Millions of chemical compounds exist. To make the study of these compounds easier, chemists divide them into a number of categories. Nearly all compounds can be classified as either organic or inorganic. Organic compounds contain the element carbon; inorganic compounds do not. A few important exceptions to that rule exist, as indicated in the description of such compounds.

Both organic and inorganic compounds can be further divided into more limited categories, sometimes called families of compounds. Some families of organic compounds are the hydrocarbons (made of carbon and hydrogen only), alcohols (containing the -OH group), and carboxylic acids (containing the -COOH groups). Many interesting and important organic compounds belong to the polymer family. Polymers consist of very large molecules in which a single small unit (called the monomer) is repeated hundreds or thousands of times over. Some polymers are made from two or, rarely, three monomers joined to each other in long chains.

Most inorganic compounds can be classified into one of four major groups. Those groups are the acids (all of which contain at least one hydrogen (H) atom), bases (which all have a hydroxide (OH) group), oxides (which all have an oxygen (O)), and salts (which include almost everything else). A few organic and inorganic compounds described in this book do not easily fit into any of these families. They are classified simply as organic or inorganic.

Molecular Weight: The molecular weight of a compound is equal to the weight of all the elements of which it is made. The molecular weight of carbon dioxide (CO_2), for example, is equal to the atomic weight of carbon (12) plus two times the atomic weight of oxygen (2 x 16 = 32), or 44. Chemists have been studying atomic weights and molecular weights for a long time, and the molecular weights of most compounds are now known with a high degree of certainty. The molecular weights expressed in this book are taken from the *Handbook of Chemistry and Physics*, 86th edition, published in 2005. The Handbook is one of the oldest, most widely used, and most highly regarded reference books in chemistry.

Melting Point and Boiling Point: The melting point of a compound is the temperature at which it changes from a solid to a liquid. Its boiling point is the temperature at which it changes from a liquid to a gas. Most organic compounds have precise melting points and/or, sometimes, precise boiling points. This fact is used to identify organic compounds. Suppose a chemist finds that a certain unknown compound melts at exactly 16.5°C. Reference books show that only a small number of compounds melt at exactly that temperature (one of which is capryllic acid, responsible for the distinctive odor of some goats). This information helps the chemist identify the unknown compound.

Inorganic compounds usually do not have such precise melting points. In fact, they may melt over a range of temperatures (from 50°C to 55°C, for example) or sublime without melting. Sublimation is the process by which a substance changes from a solid to gas without going through the liquid phase. Other inorganic compounds decompose, or break apart, when heated and do not have a true melting point.

Researchers often find different melting points and boiling points for the same compound, depending on the reference book they use. The reason for this discrepancy is that many scientists have measured the melting points and boiling points of compounds. Those scientists do not always get the same result. So, it is difficult to know what the "true" or "most correct" value is for these properties. In this book, the melting points and boiling points stated are taken from the *Handbook of Chemistry and Physics*.

Some compounds, for a variety of reasons, have no specific melting or boiling point. The term "not applicable" is used to indicate this fact.

Solubility: The solubility of a compound is its tendency to dissolve in some (usually) liquid, such as water, alcohol, or acetone. Solubility is an important property because most chemical reactions occur only when the reactants (the substances reacting with each other) are dissolved. The most common solvent for inorganic compounds is water. The most common solvents for organic compounds are the so-called organic solvents, which include alcohol, ether, acetone, and benzene. The solubility section in the entry for each compound lists the solvents in which it will dissolve well (listed as "soluble"), to a slight extent ("slightly soluble"), or not at all ("insoluble").

Overview: The overview provides a general introduction to the compound, with a pronunciation of its name, a brief history of its discovery and/or use, and other general information.

How It Is Made: This section explains how the compound is extracted from the earth or from natural materials and/or how it is made synthetically (artificially). Some production methods are difficult to describe because they include reactants (beginning compounds) with difficult chemical names not familiar to most people with little or no background in chemistry. Readers with a special interest in the synthesis (artificial production) of these compounds should consult their local librarian or a chemistry teacher at a local high school or college for references that contain more information on the process in question. The For Further Information section may also contain this information.

Interesting Facts This section contains facts and tidbits of information about compounds that may not be essential to a chemist, an inventor, or some other scientific specialist, but may be of interest to the general reader.

Common Uses and Potential Hazards Chemical compounds are often of greatest interest because of the way they can be used in medicine, industry, or some other practical application. This section lists the most important uses of each compound described in the book.

All chemical compounds pose some risk to humans. One might think that water, sugar, and salt are the safest compounds in the world. But, of course, one can drown in water, become seriously overweight by eating too much sugar, and develop heart problems by using too much salt. The risk posed by a chemical compound really depends on a number of factors, one of the most important of which is the amount of the compound to which one is exposed. The safest rule to follow in dealing with chemical compounds is that they are ALL dangerous under some circumstances. One should always avoid spilling any chemical compound on the skin, inhaling its fumes, or swallowing any of the compound. If an accident of this kind occurs, one should seek professional medical advice immediately. This book is not a substitute for prompt first aid properly applied.

Having said all that, some compounds do pose more serious health threats than others, and some individuals are at greater risks than others. Those special health risks are mentioned toward the end of the "Common Uses and Potential Hazards" section of each entry.

For Further Information As the name suggests, this section provides ideas for books, articles, and Internet sources that provide additional information on the chemical compound listed.

ADDED FEATURES

Chemical Compounds contains several features to help answer questions related to compounds, their properties, and their uses.

- The book contains three appendixes: a list by formula, list by element contained in compounds, and list by type of compound.

- Each entry contains up to two illustrations to show the relationship of the atoms in a compound to each other, one a black and white structural formula, and one a color ball-and-stick model of a molecule or portion of a molecule of the compound.

- A chronology and timeline in each volume locates significant dates in the development of chemical compounds with other historical events.

- "For Further Information," a list of useful books, periodicals, and websites, provides links to further learning opportunities.

- The comprehensive index, which appears in each volume, quickly points readers to compounds, people, and events mentioned throughout *Chemical Compounds*.

ACKNOWLEDGMENTS

In compiling this reference, the editors have been fortunate in being able to rely upon the expertise and contributions of the following educators who served as advisors:

Ruth Mormon, Media Specialist, The Meadows School, Las Vegas, Nevada

Cathy Chauvette, Sherwood Regional Library, Alexandria, Virginia

Jan Sarratt, John E. Ewing Middle School, Gaffney, South Carolina

Rachel Badanowski, Southfield High School, Southfield, Michigan

The editors would also like to thank the artists of Publishers Resource Group, under the lead of Farley Pedini, for their fast and accurate work and grace under pressure.

COMMENTS AND SUGGESTIONS

We welcome your comments on *Chemical Compounds.* Please write: Editors, *Chemical Compounds*, U•X•L, 27500 Drake Rd., Farmington Hills, MI 48331; call toll-free 1-800-877-4253; fax, 248-699-8097; or send e-mail via http://www.gale.com.

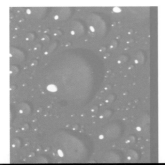

timeline of the development of chemical compounds

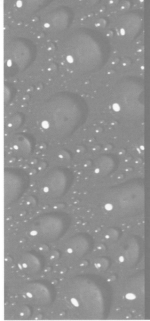

C. 3000 BCE • Egyptians develop a method for converting gypsum to plaster of Paris, which is then used as mortar to join blocks in buildings.

C. 2700 BCE • Chinese documents mention sodium chloride and the consumption of tea.

C. 1550 BCE • The analgesic properties of willow tree bark, from which salicylic acid comes, are described in Egyptian scrolls.

C. 1000 BCE • Ancient Egyptians use dried peppermint leaves.

800 BCE • Chinese and Arabic civilizations use borax for making glass and in jewelry work.

510 BCE • Persian emperor Darius makes the first recorded reference to sugar when he refers to the sugar cane growing on the banks of the Indus River.

184 BCE • Roman writer Cato the Elder describes a method of producing calcium oxide.

c. 1st century CE • Roman philosopher Pliny the Elder writes about a substance he calls hammoniacus sal, which appears to have been ammonium chloride.

1st century CE • The first recipes calling for the use of pectin to make jams and jellies are recorded.

c. 575 CE • The cultivation of the coffee tree begins in Africa.

659 • Cinnamaldehyde is described in the famous Chinese medical text, the *Tang Materia Medica.*

8th century • Arabian chemist Abu Musa Jabir ibn Hayyan, also known as Geber, writes about his work with several compounds, such as sodium chloride, sulfuric acid, nitric acid, citric acid, and acetic acid.

1242 • English natural philosopher Roger Bacon describes a method for making gunpowder.

Late 1200s • First mention of camphor by a Westerner occurs in the writings of Marco Polo.

4000 BCE			8TH CENTURY BCE	C. 6 BCE	622
Iron Age begins in Egypt.			First recorded Olympic Games.	Jesus of Nazareth is born.	Mohammed's flight from Mecca to Medina.
4000 BCE	3000 BCE	2000 BCE	1000 BCE	1 CE	500

1300s • Potassium sulfate becomes known to alchemists.

1500s • Spanish explorers bring vanilla to Europe from South and Central America, where it had already been used to flavor food.

1603 • Flemish chemist Jan Baptista van Helmont isolates a new gas produced during the combustion of wood, which is eventually called carbon dioxide.

1608 • Potash is one of the first chemicals to be exported by American colonists, with shipments leaving Jamestown, Virginia.

1610 • French alchemist Jean Béguin prepares acetone.

1620 • Flemish physician and alchemist Jan Baptista van Helmont first discovers nitric oxide.

1625 • German chemist Johann Rudolf Glauber is believed to have been the first to produce hydrogen chloride in a reasonably pure form. Later he is first to make ammonium nitrate artificially.

1695 • The term *Epsom salts* is introduced by British naturalist Nehemiah Grew, who names the compound after the spring waters near Epsom, England, from which it was often extracted.

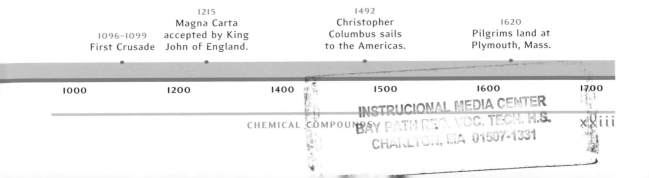

	1215		1492		1620	
	Magna Carta		Christopher		Pilgrims land at	
1096-1099	accepted by King		Columbus sails		Plymouth, Mass.	
First Crusade	John of England.		to the Americas.			

1000	1200	1400	1500	1600	1700

1700 • German chemist Georg Ernst Stahl extracts acetic acid from vinegar by distillation.

1702 • German chemist Wilhelm Homberg is believed to be the first person to prepare boric acid in Europe.

1720s • German chemist Johann Schulze makes discoveries that lead to using silver nitrate in printing and photography.

1746 • The first commercially successful method for making sulfuric acid is developed.

1747 • German chemist Andreas Sigismund Marggraf isolates a sweet substance from raisins that comes to be known as glucose.

1753 • James Lind reports that citrus fruits are the most effective means of preventing scurvy.

1769 • Oxalic acid is first isolated by German chemist Johann Christian Wiegleb.

1770s • British chemist Joseph Priestly does pioneering work with the compounds carbon dioxide, carbon monoxide, hydrogen chloride, and nitrous oxide, among others.

1770s • Swedish chemist Karl Wilhelm Scheele discovers and works with phosphoric acid, glycerol, lactic acid, and potassium bitartrate.

1726
Czar Peter the Great
of Russia dies.

1754
French and Indian
War begins in
North America.

| 1700 | 1710 | 1720 | 1730 | 1740 | 1750 |

1773 • French chemist Hilaire Marin Rouelle identifies urea as a component of urine.

Late 1700s • Commercial production of sodium bicarbonate as baking soda begins.

1776 • Carbon monoxide is first prepared synthetically by French chemist Joseph Marie François de Lassone, although he mistakenly identifies it as hydrogen.

1790 • The first patent ever issued in the United States is awarded to Samuel Hopkins for a new and better way of making pearl ash.

1794 • Ethylene is first prepared by a group of Dutch chemists.

Early 1800s • Silver iodide is first used in photography by French experimenter Louis Daguerre.

1817–1821 • French chemists Joseph Bienaimé Caventou and Pierre Joseph Pelletier successfully extract caffeine, quinine, strychnine, brucine, chinchonine, and chlorophyll from a variety of plants.

1817 • Irish pharmacist Sir James Murray uses magnesium hydroxide in water to treat stomach and other ailments. The compound is eventually called milk of magnesia.

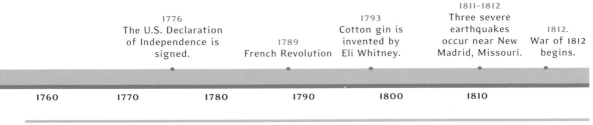

1776
The U.S. Declaration
of Independence is
signed.

1789
French Revolution

1793
Cotton gin is
invented by
Eli Whitney.

1811–1812
Three severe
earthquakes
occur near New
Madrid, Missouri.

1812.
War of 1812
begins.

1760   1770   1780   1790   1800   1810

1818 • Hydrogen peroxide is discovered by French chemist Louis Jacques Thénard.

1819 • French naturalist Henri Braconnot discovers cellulose.

1825 • British chemist and physicist Michael Faraday discovers "bicarburet of hydrogen," which is later called benzene.

1830 • Peregrine Phillips, a British vinegar merchant from England, develops the contact process for making sulfuric acid. In the early 21st century it is still the most common way to make sulfuric acid.

1831 • Chloroform is discovered almost simultaneously by American, French, and German chemists. Its use as an anesthetic is discovered in 1847.

1831 • Beta-carotene is first isolated by German chemist Heinrich Wilhelm Ferdinand Wackenroder.

1834 • Cellulose is first isolated and analyzed by French botanist Anselme Payen.

1835 • Polyvinyl chloride is first discovered accidentally by French physicist and chemist Henry Victor Regnault. PVC is rediscovered (again accidentally) in 1926.

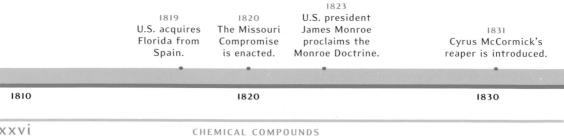

1819
U.S. acquires Florida from Spain.

1820
The Missouri Compromise is enacted.

1823
U.S. president James Monroe proclaims the Monroe Doctrine.

1831
Cyrus McCormick's reaper is introduced.

1810   1820   1830

1836 • British chemist Edmund Davy discovers acetylene.

1838 • French chemist Pierre Joseph Pelletier discovers toluene.

1839 • German-born French chemist Henri Victor Regnault first prepares carbon tetrachloride.

1839 • German druggist Eduard Simon discovers styrene in petroleum.

1845 • Swiss-German chemist Christian Friedrich Schönbein discovers cellulose nitrate.

1846 • Americans Austin Church and John Dwight form a company to make and sell sodium bicarbonate. The product will become known as Arm & Hammer® baking soda.

Mid 1800s • Hydrogen peroxide is first used commercially—primarily to bleach hats.

1850s • Oil is first discovered in the United States in western Pennsylvania.

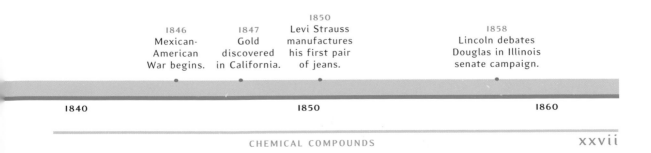

| 1846 Mexican-American War begins. | 1847 Gold discovered in California. | 1850 Levi Strauss manufactures his first pair of jeans. | 1858 Lincoln debates Douglas in Illinois senate campaign. |

1840   1850   1860

1853 • French chemist Charles Frederick Gerhardt develops a method for reacting salicylic acid (the active ingredient in salicin) with acetic acid to make the first primitive form of aspirin.

1859 • Ethylene glycol and ethylene oxide are first prepared by French chemist Charles Adolphe Wurtz.

1860s • Swedish chemist Alfred Nobel develops a process for manufacturing nitroglycerin on a large scale.

1863 • TNT is discovered by German chemist Joseph Wilbrand, although the compound is not recognized as an explosive until nearly 30 years later.

1865 • The use of carbolic acid as an antiseptic is first suggested by Sir Joseph Lister.

1865 • German botanist Julius von Sachs demonstrates that chlorophyll is responsible for photosynthetic reactions that take place within the cells of leaves.

1870 • American chemist Robert Augustus Chesebrough extracts and purifies petrolatum from petroleum and begins manufacturing it, eventually using the name Vaseline™.

1873 • German chemist Harmon Northrop Morse rediscovers and synthesizes acetaminophen. It had been discovered originally in 1852, but at the time it was ignored.

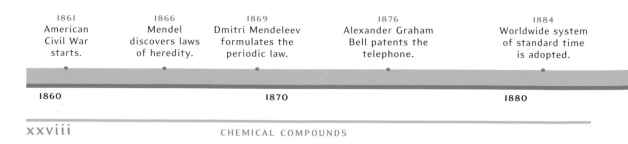

1861	1866	1869	1876	1884
American Civil War starts.	Mendel discovers laws of heredity.	Dmitri Mendeleev formulates the periodic law.	Alexander Graham Bell patents the telephone.	Worldwide system of standard time is adopted.

1860   1870   1880

1879 • Saccharin, the first artificial sweetener discovered, is synthesized accidentally by Johns Hopkins researchers Constantine Fahlberg and Ira Remsen.

1879 • Riboflavin is first observed by British chemist Alexander Wynter Blyth.

1883 • Copper(I) oxide is the first substance found to have semiconducting properties.

1886 • American chemist Charles Martin Hall invents a method for making aluminum metal from aluminum oxide, which drastically cuts the price of aluminum.

1889 • French physiologist Charles E. Brown-Séquard performs early experiments on the effects of testosterone.

1890s • Commercial production of perchlorates begins.

1890s–early 1900s • British chemists Charles Frederick Cross, Edward John Bevan, and Clayton Beadle identify the compound now known as cellulose. They also develop rayon.

Late 1890s • Artificial methods of the production of pure vanillin are developed.

1901 • The effects of fluorides in preventing tooth decay are first observed.

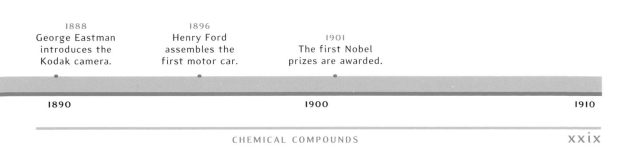

1888
George Eastman
introduces the
Kodak camera.

1896
Henry Ford
assembles the
first motor car.

1901
The first Nobel
prizes are awarded.

1890   1900   1910

1937 • German forensic scientist Walter Specht discovers that blood can act as the catalyst needed to produce chemiluminescence with luminol, a compound discovered in the late 1800s.

1937 • Plexiglas® (made from polymethylmethacrylate) is exhibited at the World's Trade Fair in Paris.

1937 • The basic process for making polyurethanes is first developed by German chemist Otto Bayer.

1937 • The cyclamate family of compounds is discovered by Michael Sveda, a graduate student at the University of Illinois.

1938 • Polytetrafluoroethylene is invented by Roy J. Plunkett by accident at DuPont's Jackson Laboratory.

1939 • Swiss chemist Paul Hermann Müller finds that DDT is very effective as an insecticide, which makes it useful in preventing infectious diseases such as malaria.

1939-1945 • During World War II, the U.S. military finds a number of uses for nylon, polyurethanes, polystyrene, percholorates, and silica gel.

Early 1940s • Penicillin is first produced for human use and is valuable in saving the lives of soldiers wounded in World War II.

1939
World War II
begins.

1941
First regular
television
broadcasts
begin.

1942
Irving Berlin
writes song
"White
Christmas."

1945
U.S. drops
two atomic
bombs
on Japan.

1946
First
"baby boomers"
are born.

1935   1940   1945

CHEMICAL COMPOUNDS

1940s • A research chemist at the General Electric Company, E. G. Rochow, finds an efficient way of making organosiloxanes in large quantities.

1941 • Folic acid is isolated and identified by American researcher Henry K. Mitchell.

1941 • The first polyurethane adhesive, for the joining of rubber and glass, is made.

1942 • American researchers Harry Coover and Fred Joiner discover cyanoacrylate.

1946 • DEET is patented by the U.S. Army for use on military personnel working in insect-infested areas. It is made available to the public in 1957.

1947 • On April 16, an ammonium nitrate explosion in Texas City, Texas, becomes the worst industrial accident in U.S.

1950s • Earliest reports surface about athletes using testosterone to enhance their sports performance.

1950s • A stretchable material made of polyurethane, called spandex, is introduced.

1951 • Phillips Petroleum Company begins selling polypropylene under the trade name of Marlex®.

1953 • Polycarbonate, polyethylene, and synthetic rubber are developed.

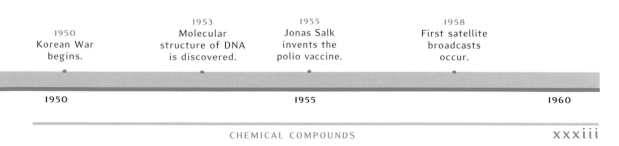

| 1950 Korean War begins. | 1953 Molecular structure of DNA is discovered. | 1955 Jonas Salk invents the polio vaccine. | 1958 First satellite broadcasts occur. |

1950   1955   1960

1955 • Proctor & Gamble releases the first toothpaste containing stannous fluoride, Crest®.

Mid 1950s • Wham-O creates the hula-hoop—a ring of plastic that is made with low-grade polyethylene.

1956 • British chemist Dorothy Hodgkin determines the chemical structure of cyanocobalamin.

1958 • Scientist W. Barnes of the chemical firm T. & H. Smith in Edinburgh, Scotland, discovers denatonium benzoate.

Early 1960s • Ibuprofen is developed by researchers at the Boots Company, a British drug manufacturer.

1960s • Triclosan becomes a common ingredient in soaps and other cleaning projects.

1960s • MTBE is first synthesized by researchers at the Atlantic Richfield Corporation as an additive designed to increase the fuel efficiency of gasoline.

1962 • Amoxicillin is discovered by researchers at the Beecham pharmaceutical laboratories.

1965 • Aspartame is discovered accidentally by James M. Schlatter.

1963
U.S. President
John F. Kennedy
is assassinated.

1975
Vietnam War
ends.

1950   1960   1970

1980s • A ceramic form of copper(I) oxide is found to have superconducting properties at temperatures higher than previously known superconductors.

1980s • Polycarbonate bottles begin to replace the more cumbersome and breakable glass bottles.

1987 • Procter & Gamble seeks FDA approval of sucrose polyester. Ten years pass before the FDA grants that approval.

1994 • The U.S. Food and Drug Administration approves the sale of naproxen as an over-the-counter medication.

1995 • On April 19, American citizens Timothy McVeigh and Terry Nichols use a truckload of ammonium nitrate and other materials to blow up the Alfred P. Murrah Federal Building in Oklahoma City, Oklahoma.

Early 2000s • Some 350,000 propane-powered vehicles exist in the United States and about 4 million are used worldwide.

2004 • The leading chemical compound manufactured in the United States is sulfuric acid, with 37,515,000 metric tons (41,266,000 short tons) produced. Next is ethylene, with about 26.7 million metric tons (29.4 million short tons) produced.

1981
First personal computers become available.

1989
The oil tanker *Exxon Valdez* sinks off Alaska.

1991
Soviet Union is dissolved.

2001
World Trade Center in New York City is destroyed.

1980   1990   2000   2006

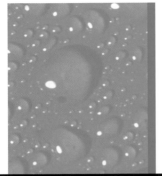

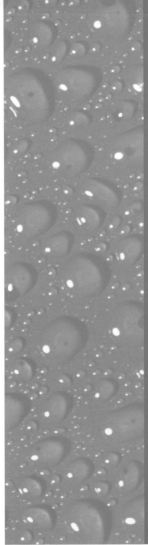

words to know

ACETYL The organic group of acetic acid.

ADHESIVE A substance used to bond two surfaces together.

ALCHEMY An ancient field of study from which the modern science of chemistry evolved.

ALKALI A chemical base that can combine with an acid to produce a salt.

ALKALINE A substance that has a pH higher than 7.

ALKALOID An organic base that contains the element nitrogen.

ALKANE A type of hydrocarbon that has no double bonds because it contains the maximum possible number of hydrogen

ALKENE A kind of hydrocarbon with at least one double bond between carbons.

ALKYL GROUP A chemical group containing hydrogen and carbon atoms.

ALLOTROPE A form of an element that is different from its typical form, with a different chemical bond structure between atoms.

AMIDE An organic compound that includes the CON group bound to hydrogen.

AMINO ACID An organic compound that contains at least one carboxyl group (-COOH) and one amino group (-NH$_2$). They are the building blocks of which proteins are made.

ANALGESIC A substance the relieves pain.

ANHYDROUS Free from water and especially water that is chemically combined in a crystalline substance.

ANION A negatively charged ion.

ANODE The electrode in a battery in which electrons are lost (oxidized).

AROMATIC COMPOUND A compound whose chemical structure is based on that of benzene (C$_6$H$_6$).

BIODEGRADABLE Something that can be easily broken down by the action of bacteria.

BLOCK COPOLYMER A polymer composed of two or more different polymers, each of which clumps in blocks of identical molecules.

BORATE A salt that contains boron.

BRINE Salt water; water with a large amount of salt dissolved in it, such as seawater or water used to pickle vegetables.

BYPRODUCT A product that is made while making something else.

CARBOHYDRATES Organic compounds composed of carbon, oxygen, and hydrogen, which are used by the body as food.

CARBOXYL GROUPS Groups of atoms consisting of a carbon atom double bonded to an oxygen atom and single bonded to a hydroxyl (-OH) group (-COOH).

CARCINOGEN A substance that causes cancer in humans or other animals.

CATALYST A material that increases the rate of a chemical reaction without undergoing any change in its own chemical structure.

CATHODE The electrode in a battery through which electrons enter the fuel cell.

CATION A positively charged ion.

CAUSTIC Capable of burning or eating away, usually by the action of chemical reactions.

CENTRIFUGE A device that separates substances that have different densities by using centrifugal force.

CHELATE A chemical compound that is in the form of a ring. It usually contains one metal ion attached to a minimum of two nonmetal ions by coordinate bonds.

CHEMILUMINESCENCE Light produced by a chemical reaction.

CHIRAL A molecule with different left-handed and right-handed forms; not mirror symmetric.

CHLOROFLUOROCARBONS (CFCS) A family of chemicals made up of carbon, chlorine, and fluorine. CFCs were used as a refrigerant and propellant before they were banned for fear that they were destroying the ozone layer.

CHROMATOGRAPHY A process by which a mixture of substances passes through a column consisting of some material that causes the individual components in the mixture to separate from each other.

COAGULATE To make a liquid become a semisolid.

COENZYME A chemical compound that works along with an enzyme to increase the rate at which chemical reactions take place.

COMPOUND A substance formed of two or more elements that are chemically combined.

COPOLYMER A polymer made from more than one type of monomer.

COVALENT COMPOUND A compound in which the atoms are bonded to each other by sharing electrons.

CROSS-LINKED Polymer chains that are linked together to create a chemical bond.

CRYOGENICS The study of substances at very low temperatures, using substances such as liquefied hydrogen or liquefied helium.

DECOMPOSE To break a substance down into its most basic elements.

DENATURED To be made not fit for drinking.

DERIVATIVE Something gotten or received from another source.

DESICCANT Chemical agent that absorbs or adsorbs moisture.

DIATOMIC Composed of two atoms.

DISACCHARIDE A compound formed by the joining of two sugar molecules.

DISPERSANT A substance that keeps another substance from clumping together or becoming lumpy.

DISTILLATION A process of separating liquid by heating it and then condensing its vapor.

ELASTOMER A polymer known for its flexibility and elastic qualities; a type of rubber.

ELECTRODE A conductor through which an electric current flows.

ELECTROLYSIS A process in which an electric current is used to bring about chemical changes.

ELECTROLYTE A substance which, when dissolved in water, will conduct an electric current.

EMULSIFIER A substance that combines two other substances together that do not usually mix together and ensures they are spread evenly.

ESTER A compound formed by the reaction between an acid and an alcohol.

EXOTHERMIC Accompanied by the freeing of heat.

FAT An ester formed in the reaction between glycerol $(C_3H_5(OH)_3)$ and a fatty acid, an organic acid with more than eight carbon atoms.

FERROUS Containing or made from iron.

FLOCCULANT A type of polymer, or large man-made particle, that is created by a repetitive chain of atoms.

FLUOROCARBON A chemical compound that contains carbon and fluorine, such as chlorofluorocarbon.

FOSSIL FUEL A fuel, such as petroleum, natural gas, or coal, formed from the compression of plant and animal matter underground millions of years ago.

FREE RADICAL An atom or group of atoms with a single unpaired electron that can damage healthy cells in the body.

G/MOL Grams per mole: a measure of molar mass that indicates the amount of the compound that is found in a mole of the compound. The molecular weight (also known as relative molecular mass) is the same number as the molar mass, but the g/mol unit is not used.

GLOBAL WARMING The increase in the average global temperature.

GLUCONEOGENESIS The production of glucose from non-carbohydrate sources, such as proteins and fats.

GLYCOGEN A carbohydrate that is stored in the liver and muscles, which breaks down into glucose.

GREENHOUSE EFFECT The increase of the average global temperature due to the trapping of heat in the atmosphere.

HARD WATER Water with a high mineral content that does not lather easily.

HELIX A spiral; a common shape for protein molecules.

HORMONE A chemical that delivers messages from one cell or organ to another.

HYDRATE A chemical compound formed when one or more molecules of water is added physically to the molecule of some other substance.

HYDROCARBON A chemical compound consisting of only carbon and hydrogen, such as fossil fuels.

HYDROGENATION A chemical reaction of a substance with molecular hydrogen, usually in the presence of a catalyst.

HYDROLYSIS The process by which a compound reacts with water to form two new compounds.

INCOMPLETE COMBUSTION Combustion that occurs in such a way that fuel is not completely oxidized.

INERT A substance that is chemically inactive.

INORGANIC Relating to or obtained from nonliving things.

ION An atom or molecule with an electrical charge, either positive or negative.

IONIC BOND A force that attracts and holds positive and negative ions together.

IONIC COMPOUND A compound that is composed of positive and negative ions, so that the total charge of the positive ions is balanced by the total charge of the negative ions.

ISOMER Two or more forms of a chemical compound with the same molecular formula, but different structural formulas and different chemical and physical properties.

ISOTOPE A form of an element with the usual number of protons in the nucleus but more or less than the usual number of electrons.

KREBS CYCLE A series of chemical reactions in the body that form part of the pathway by which cells break down carbohydrates, fats, and proteins for energy. Also called the citric acid cycle.

LATENT Lying hidden or undeveloped.

LEACH Passing a liquid through something else in order to dissolve minerals from it.

LEWIS ACID An acid that can accept two electrons and form a coordinate covalent bond.

LIPID An organic compound that is insoluble in water, but soluble in most organic solvents, such as alcohol, ether, and acetone.

MEGATON A unit of explosive force equal to one million metric tons of TNT.

METABOLISM All of the chemical reactions that occur in cells by which fats, carbohydrates, and other compounds are broken down to produce energy and the compounds needed to build new cells and tissues.

METALLURGY The science of working with metals and ores.

METHYLXANTHINE A family of chemicals including caffeine, theobromine, and theophylline, many of which are stimulants.

MINERALOGIST A scientist who studies minerals.

MISCIBLE Able to be mixed; especially applies to the mixing of one liquid with another.

MIXTURE A collection of two or more elements or compounds with no definite composition.

MONOMER A single molecule that can be strung together with like molecules to form a polymer.

MONOSACCHARIDE A simple sugar, made up of three to nine carbon atoms.

MORDANT A substance used in dyeing and printing that reacts chemically with both a dye and the material being dyed to help hold the dye permanently to the material.

NARCOTIC An addictive drug that relieves pain and causes drowsiness.

NEUROTRANSMITTER A chemical that relays signals along neurons.

NEUTRALIZE To make a substance neutral that is neither acidic or alkaline.

NITRATING AGENT A substance that turns other substances into nitrates, which are compounds containing NO_3.

NSAID Non-steroidal anti-inflammatory drug, a drug that can stop pain and prevent inflammation and fever but that is not a steroid and does not have the same side effects as steroids.

ORGANIC Relating to or obtained from living things. In chemistry, refers to compounds made of carbon combined with other elements.

OXIDANT A substance that causes oxidation of a compound by removing electrodes from the compound. Also known as oxidizing agent.

OXIDATION STATE The sum of negative and positive charges, which indirectly indicates the number of electrons accepted or donated in the bond between elements.

OXIDIZE To combine a substance with oxygen, or to remove hydrogen from a molecule using oxygen, or to remove electrons from a molecule.

PARTICULATE MATTER Tiny particles of pollutants suspended in the air.

PETROCHEMICALS Chemical compounds that form in rocks, such as petroleum and coal.

PH (POTENTIAL HYDROGEN) The acidity or alkalinity of a substance based on its concentration of hydrogen ions.

PHOSPHATE A compound that is a salt of phosphoric acid, which consists of phosphorus, oxygen, sometimes hydrogen, and another element or ion.

PHOTOELECTRIC EFFECT The emission of electrons by a substance, especially metal, when light falls on its surface.

PHOTOSYNTHESIS The process by which green plants and some other organisms using the energy in sunlight to convert carbon dioxide and water into carbohydrates and oxygen.

PHOTOVOLTAIC EFFECT A type of photoelectric effect where light is converted to electrical voltage by a substance.

PLASTICIZER A substance added to plastics to make them stronger and more flexible.

POLYAMIDE A polymer, such as nylon, containing recurrent amide groups linking segments of the polymer chain.

POLYMER A substance composed of very large molecules built up by linking small molecules over and over again.

POLYSACCHARIDE A very large molecule made of many thousands of simple sugar molecules joined to each other in long, complex chains.

PRECIPITATE A solid material that settles out of a solution, often as the result of a chemical reaction.

PRECURSOR A compound that gives rise to some other compound in a series of reactions.

PROPRIETARY Manufactured, sold, or known only by the owner of the item's patent.

PROSTAGLANDINS A group of potent hormone-like substances that are produced in various tissues of the body. Prostaglandins help with a wide range of physiological functions, such as control of blood pressure, contraction of smooth muscles, and modulation of inflammation.

PROTEIN A large, complex compound made of long chains of amino acids. Proteins have a number of essential functions in living organisms.

QUARRY An open pit, often big, that is used to obtain stone.

REAGENT A substance that is employed to react, measure, or detect with another substance.

REDUCTION A chemical reaction in which oxygen is removed from a substance or electrons are added to a substance.

REFRACTORY A material with a high melting point, resistant to melting, often used to line the interior of industrial furnaces.

RESIN A solid or semi-solid organic material that is used to make lacquers, adhesives, plastics, and many other clear substances.

S

SALT An ionic compound where the anion is derived from an acid.

SEMICONDUCTOR A material that has an electrical conductance between that of an insulator and a conductor. When charged with electricity or light, semiconductors change their state from nonconductive to conductive or vice versa.

SEQUESTERING AGENT A substance that binds to metals in water to prevent them from combining with other components in the water and forming compounds that could stain (sequestering agents are sometimes used in cleaning products). Also called a "chelating agent."

SILICATE A salt in which the anion contains both oxygen and silicon.

SOLUBLE Capable of being dissolved in a liquid such as water.

SOLUTION A mixture whose properties and uniform throughout the mixture sample.

SOLVENT A substance that is able to dissolve one or more other substances.

SUBLIME To go from solid to gaseous form without passing through a liquid phase.

SUPERCONDUCTIVITY A state in which a material loses all electrical resistance: once established, an electrical current will flow forever.

SUPERCOOLED WATER Water that remains in liquid form, even though its temperature is below 0°C.

SUSPENSION A mixture of two substances that do not dissolve in each other.

SYNTHESIZE To produce a chemical by combining simpler chemicals.

TERATOGENIC Causing birth defects; comes from the Greek word "teratogenesis," meaning "monster-making."

THERMAL Involving the use of heat.

TOXIC Poisonous or acting like a poison.

TOXIN A poison, usually produced by microorganisms or by plants or animals.

TRAP A reservoir or area within Earth's crust made of non-porous rock that can contain liquids or gases, such as water, petroleum, and natural gas.

ULTRAVIOLET LIGHT Light that is shorter in wavelength than visible light and can fade paint finishes, fabrics, and other exposed surfaces.

VASODILATOR A chemical that makes blood vessels widen, reducing blood pressure.

VISCOUS Having a syrupy quality causing a material to flow slowly.

VITRIFICATION The process by which something is changed into glass or a glassy substance, usually by heat.

VOLATILE Able to turn to vapor easily at a relatively low temperature.

WATER OF HYDRATION Water that has combined with a compound by some physical means.

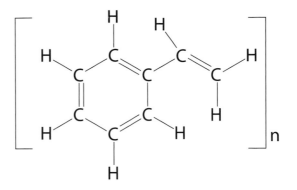

OTHER NAMES:
Styrofoam

FORMULA:
$-[-CH_2C_6H_5-]-_n$

ELEMENTS:
Carbon, hydrogen

COMPOUND TYPE:
Organic polymer

STATE:
Solid

MOLECULAR WEIGHT:
Varies g/mol

MELTING POINT:
Varies widely; ranges
from 190°C–260°C
(370°F–500°F)

BOILING POINT:
Not applicable

SOLUBILITY:
Insoluble in water
and inorganic acids
and bases; soluble
in many organic
solvents, including
ethylbenzene,
chloroform, carbon
tetrachloride, and
tetrahydrofuran

KEY FACTS

Polystyrene

OVERVIEW

Polystyrene (pol-ee-STYE-reen) is a thermoplastic polymer made from styrene. A thermoplastic polymer is a material that can be repeatedly softened and hardened by alternately heating and cooling. Styrene is a hydrocarbon derived from petroleum with the formula $C_6H_5CH=CH_2$. The presence of the double bond in the styrene molecule makes it possible for styrene molecules to react with each other in long chains that constitute the polymer polystyrene.

Polystyrene is a hard, strong, transparent solid highly resistant to mechanical impact. It is an excellent thermal (heat) and electrical insulator, is easily shaped and molded in the liquid state, and takes dyes readily. It can be produced in a wide variety of shapes and forms, including sheets, plates, rods, beads, and foams.

The history of polystyrene dates to 1839 when a German apothecary (druggist) named Eduard Simon discovered styrene in petroleum. Later scientists attempted to incorporate

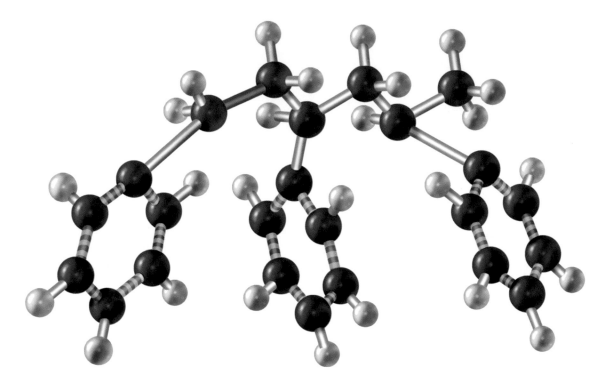

Polystyrene. White atoms are hydrogen and black atoms are carbon. Gray sticks indicate double bonds. Striped sticks indicate benzene rings.
PUBLISHERS RESOURCE GROUP

styrene into some of the commercial products they made, such as rubber tires. But a polymer based on styrene was not actually produced until 1930 when researchers at the German chemical firm of I. G. Farben discovered how to make the material. Farber's parent company, BASF, shortly made the product commercially available and in 1937, Dow Chemical first made the compound available in the United States. During World War II, polystyrene was used for the manufacture of synthetic rubber products. After the war, the number of commercial and industrial uses expanded rapidly. Today, it is virtually impossible to avoid polystyrene products in one's daily life.

HOW IT IS MADE

Compounds like styrene with double bonds often polymerize easily. The double bond on a styrene molecule breaks open and a hydrogen atom from a second styrene molecule adds to one side of the double bond, while the rest of the second styrene molecule adds to the second side of the double bond. The product of this reaction still has a double bond.

Interesting Facts

- Polystyrene is sold commercially under more than a hundred trade names, the most famous of which is probably Styrofoam®.

- Only about 5 percent of a styrofoam cup is polystyrene. The rest is air.

- One of the innovative uses for polystyrene is as a building material for the construction of new houses. Scientists suggest that it is perfect for the purpose: lightweight, inexpensive, strong, a good insulator, and available all over the world. One of the first applications suggested for polystyrene as a building material is in the construction of houses in Afghanistan, where many families have lost their homes after two decades of wars and earthquakes.

So the reaction can be repeated a second time; and a third time; and a fourth time; and so on. One goal of research on polystyrene has been to determine how the size of the polystyrene affects its properties (and, therefore, its uses) and how to stop the polymerization reaction at some desired point.

All that is needed to start the polymerization of styrene is a material that will cause the first double bond to break. Such materials are known as polymerization initiators. One of the most common initiators used in the polymerization of styrene is benzoyl peroxide ($C_6H_5COOOCOC_6H_5$). Once the polymerization reaction begins, it tends to release enough energy for the reaction to continue on its own.

An especially popular form of polystyrene is known as expanded polystyrene. It is made by blending air with molten polystyrene to make a lightweight foam sold under the trade name of Styrofoam®.

COMMON USES AND POTENTIAL HAZARDS

Polystyrene is the fourth largest thermoplastic polymer made in the United States by production volume. It is used in the manufacture of hundreds of commercial, industrial, household, and personal articles. Some examples include:

- Plastic model kits and toys;
- Containers with lids; disposable cups, plates, knives, forks, and spoons;
- "Jewel" cases for compact discs and cases for audiocassettes;
- Plastic coat hangers and plastic trays;
- Refrigerator doors and air conditioner cases;
- Housing for machines; and
- Cabinets for clocks, radios, and television sets.

Some uses of expanded polystyrene include:

- In all kinds of containers to keep foods either hot or cold (such as ice chests);
- Egg cartons;
- Fillers in shipping containers;
- Packages for carry-out foods;
- Insulation for buildings;
- In the construction of boats; and
- For the construction of some types of furniture.

Polystyrene dust and powder formed during production can be a mild irritant to the eyes, skin, and respiratory system. But even for workers in the field, the risk is regarded as being very low. A more serious problem posed by the compound is the risk it poses for the environment. About half of all the polystyrene produced in the United States is used for packaging and "one-time use" purposes. That is, someone uses the product and then throws it away. Since polystyrene does not readily decompose, it tends to accumulate in landfills and dumps. Some environmentalists point out that large volumes of discarded polystyrene contribute significantly to the nation's solid waste disposal problems. Industry spokespersons, however, point out that polystyrene accounts for less than one percent of all solid wastes. In any case, a number of industries and companies have attempted to reduce the amount of polystyrene used in their products in order to

Words to Know

HYDROCARBON a compound consisting of carbon and hydrogen.

POLYMER a compound consisting of very large molecules made of one or two small repeated units called monomers.

THERMOPLASTIC can be repeatedly softened and hardened by alternately heating and cooling.

cut back on their contribution to the solid waste disposal problem.

FOR FURTHER INFORMATION

Boyd, Clark. "Polystyrene homes planned for Afghans." BBC News. http://news.bbc.co.uk/2/hi/technology/3528716.stm (accessed on October 26, 2005).

"Energy & Waste—Landfilling." Energy Information Administration, Department of Energy. http://www.eia.doe.gov/kids/energyfacts/saving/recycling/solidwaste/landfiller.html (accessed on October 26, 2005).

"Polystyrene." U.S. Environmental Protection Agency, Emissions Factors and Policy Applications Center. http://www.epa.gov/ttn/chief/ap42/ch06/final/c06s06-3.pdf (accessed on October 26, 2005).

"Polystyrene Packaging Delivers!" Polystyrene Packaging Council. http://www.polystyrene.org/ (accessed on October 26, 2005).

Sims, Judith. "Polystyrene." In *Environmental Encyclopedia.* 3rd ed. Edited by Marci Bortman and Peter Brimblecombe. Detroit, Mich.: Gale, 2003.

See Also Styrene

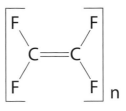

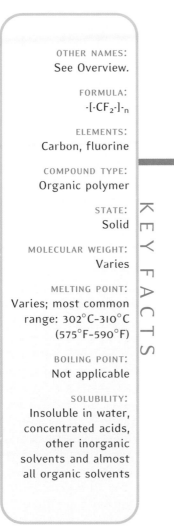

OTHER NAMES:
See Overview.

FORMULA:
-[-CF$_2$-]-$_n$

ELEMENTS:
Carbon, fluorine

COMPOUND TYPE:
Organic polymer

STATE:
Solid

MOLECULAR WEIGHT:
Varies

MELTING POINT:
Varies; most common
range: 302°C–310°C
(575°F–590°F)

BOILING POINT:
Not applicable

SOLUBILITY:
Insoluble in water,
concentrated acids,
other inorganic
solvents and almost
all organic solvents

K E Y   F A C T S

Polytetrafluoro-
ethylene

OVERVIEW

Polytetrafluoroethylene (POL-ee-tet-ruh-FLUR-oh-ETH-eh-leen) is also known as polytetrafluoroethene, tetrafluoroethy-lene polymer, PTFE, and Teflon®. Polytetrafluoroethylenes are thermoplastic polymers made from the monomer tetra-fluoroethylene (CF$_2$=CF$_2$). A thermoplastic polymer is a material that can be repeatedly softened and hardened by alternately heating and cooling. The polytetrafluoroethylenes are well known by the trade name of Teflon® although they are also available under more than a hundred other trade names, including Aflon®, Algloflon®, Ethicon®, Fluon®, Ftorlon®, Halon®, Molykote®, Polyflon®, Polytef®, and PTFE®. DuPont Chemical, one of the two largest manufacturers of polytetra-fluoroethylene, makes at least ten grades of Teflon®. These products differ from each other in the physical form in which they are provided (powders, aqueous dispersions, yarn, or film, for example) and size (molecular weight) of the product (ranging from low-molecular weight to high-molecular weight).

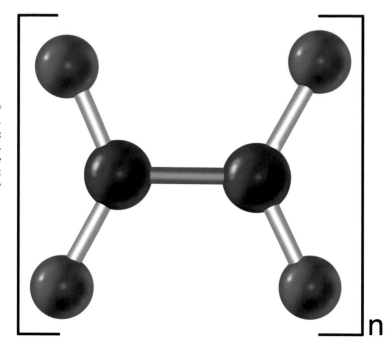

Polytetrafluoroethylene.
Black atoms are carbon;
turquoise atoms are fluorine.
Gray stick indicates double
bond. PUBLISHERS RESOURCE
GROUP

Polytetrafluoroethylene has the lowest coefficient of friction of any known substance. The coefficient of friction is a measure of how easily one substance slides over the surface of a second substance. Polytetrafluoroethylene's low coefficient of friction means that nothing will stick very well to its surface, accounting for Teflon®'s primary use in the manufacture of non-stick products.

Polytetrafluoroethylene was invented in 1938 by Roy J. Plunkett (1910-1994) quite by accident. As a research chemist at DuPont's Jackson Laboratory, Plunkett was studying compounds that might be used for refrigerants. He kept the compounds in steel tanks and was surprised on one occasion to find that the gas he wanted did not leave the storage tank when the valve was opened. He cut the tank open to see what had happened to the gas and found a waxy white material. Upon analysis, the material turned out to be polytetrafluoroethylene. The gas stored in the tank, the potential refrigerant, was tetrafluoroethylene. It had undergone polymerization spontaneously within the tank, making it possible for Plunkett to discover one of the most remarkable synthetic products in the world.

Interesting Facts

- Because Teflon® does not stick to anything else, cookware manufacturers must use a special process to get it to stay on pots and pans. They sometimes begin by blasting the pan with sand or grit to roughen the surface. Then they apply a special primer that makes the Teflon® adhere to the pan's surface.

- Teflon®'s non-stick quality has been compared to trying to make one piece of ice stick to another piece of ice.

HOW IT IS MADE

Molecules of tetrafluoroethylene contain double bonds. Any compound with double bonds has the ability to form polymers. Polymerization of tetrafluoroethylene occurs when the double bond in one molecule breaks apart. A fluorine atom from a second molecule of the monomer then adds on to one end of the broken double bond. The rest of the second molecule adds to the other end of the broken double bond. A "double-molecule," consisting of two monomers joined to each other forms: $CF_2=CF_2 + CF_2=CF_2 \rightarrow CF_3CF_2CF_2=CF_2$. The product of this reaction also contains a double bond. So the process can be repeated to form another product consisting of three monomer molecules: $CF_3CF_2CF_2=CF_2 + CF_2=CF_2 \rightarrow CF_3CF_2CF_2CF_2CF_2=CF_2$ The process is repeated hundreds or thousands of times producing a long chain of monomers with the general formula $-[-CF_2-]-_n$.

COMMON USES AND POTENTIAL HAZARDS

Perhaps the best known application of polytetrafluoroethylene is in kitchen utensils with non-stick coatings, such as pots, pans, and spatulas. Polytetrafluoroethylene is also used to coat fibers to make them water-repellant and stain-resistant. Water will bead up and roll off the surface of clothing and other materials coated with polytetrafluoroethylene instead of penetrating the fabric and possibly leaving a stain. Polytetrafluoroethylene is available as a spray treatment for carpets and furniture, forming a molecular shield to prevent water or oil-based stains from penetrating the material. Some carpets come pre-treated with a polytetrafluoroethylene product to keep them clean and fresh. The compound can also be used on wood and plastic flooring to protect it from dirt, stains, and moisure.

Automobile manufacturers used polytetrafluoroethylene in a variety of ways. Windshield wiper blades coated with polytetrafluoroethylene are smoother and stronger than uncoated blades. Automobile paint can be coated with polytetrafluoroethylene to protect a car's finish from tree sap, insects, and other residues. Automobile upholstery is often treated with polytetrafluoroethylene to protect against stains caused by spilled drinks and dirty shoes. Polytetrafluoroethylene added to oil makes it flow through an engine more smoothly, reducing wear and tear on the engine.

Polytetrafluoroethylene is used widely for a number of industrial applications. Some armor-piercing bullets are coated with the compound to reduce friction when the bullet leaves the gun barrel and increases the ease with which it breaks through armor. Many electrical cables are insulated with polytetrafluoroethylene, which is not combustible or conductive. Food processing equipment made with polytetrafluoroethylene is easier to clean and more efficient for cooking and baking than non-polytetrafluoroethylene equipment. Industrial bakers often use equipment coated with some type of polytetrafluoroethylene. The product can also be used to coat stainless steel, carbon steel, aluminum, steel alloys, brass, magnesium, glass, fiberglass, plastics, and rubber. Outdoor signs are sometimes coated with polytetrafluoroethylene to make them last longer and resist stains.

Polytetrafluoroethylene has long been regarded as an essentially safe compound with no known health effects on humans or experimental animals. Recently, questions have been raised about possible health hazards of one of the compounds used in the manufacture of polytetrafluoroethylene, perfluorooctanoic acid (PFOA). Some studies suggest that PFOA may be responsible for birth defects and the development of cancer in people who have been exposed to the chemical. Other studies show that 96 percent of the children tested in 23 states and the District of Columbia in 2001 had detectable levels of PFOA in their blood. Federal

Words to Know

POLYMER A compound consisting of very large molecules made of one or two small repeated units called monomers.

THERMOPLASTIC Able to be repeatedly softened and hardened by alternately heating and cooling.

agencies have not yet confirmed the level of risk that PFOA poses, if any, and have not listed any other chemicals used in the manufacture of polytetrafluoroethylene as hazardous to human health.

FOR FURTHER INFORMATION

"PTFE Specifications." Boedeker Plastics. http://www.boedeker.com/ptfe_p.htm (accessed on October 26, 2005).

Summer, Chris. "Teflon's Sticky Situation." BBC News Online (October 7, 2004). Available online at http://news.bbc.co.uk/1/hi/magazine/3697324.stm (accessed on October 26, 2005).

"Technical Information: Teflon® Fluorocarbon Resin." Omega. Stamford, Conn.: Omega Engineering, Inc. 2000. Available online at http://www.omega.com/pdf/tubing/technical_section/teflon_flourocarbon.asp (accessed on October 26, 2005).

"Teflon (PTFE: polytetrafluoroethylene)." Fluoride Action Network. http://www.fluoridealert.org/pesticides/polytetrafluoroethylen-page.htm (accessed on October 26, 2005).

See Also Polyethylene

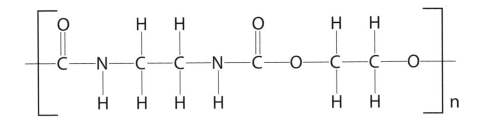

OTHER NAMES:
None

FORMULA:
-[-CONH-C$_6$H$_4$-NCOO-CH$_2$CH$_2$-O-]-$_n$; other structures are possible

ELEMENTS:
Carbon, hydrogen, oxygen, nitrogen

COMPOUND TYPE:
Organic polymer

STATE:
Solid

MOLECULAR WEIGHT:
Varies; very large

MELTING POINT:
Variable

BOILING POINT:
Not applicable

SOLUBILITY:
Insoluble in water; soluble in aromatic hydrocarbons, such as benzene and toluene

KEY FACTS

Polyurethane

OVERVIEW

Polyurethanes (pol-ee-YUR-eth-anes) are a group of thermoplastic polymers formed in the reaction between a diisocyanate and a polyol, an alcohol with two or more hydoxyl (-OH) groups. Diisocyanates are compounds that contain two isocyanate (-N=C=O) groups.

Polyurethanes are available in a variety of forms, including fibers, foams, coatings, and elastomers, rubber-like materials. Each form of polyurethane has its own set of physical and chemical properties. For example, fibers are moisture proof, stretchable, and resistant to the flow of electric current. Foams can be either rigid or flexible, with densities as low as 32 kilograms per cubic meter (2 pounds per cubic foot) to as high as 800 kilograms per cubic meter (50 pounds per cubic foot). They are excellent thermal (heat) insulators. Polyurethane coatings are hard, glossy, flexible, readily adhesive to other surfaces, and resistant to abrasion, weathering, and most inorganic chemicals. Elastomeric polyurethanes are resistant to abrasion, weathering, and most organic solvents.

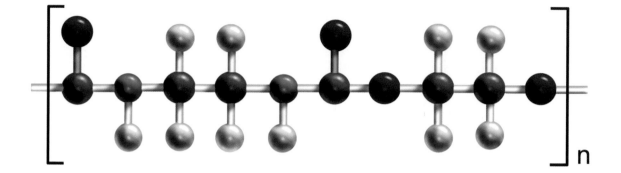

n

Polyurethane. Red atoms are oxygen; white atoms are hydrogen; black atoms are carbon; and blue atoms are nitrogen. Gray sticks show double bonds. PUBLISHERS RESOURCE GROUP

With this variety of properties, polyurethanes have a very wide range of uses.

The basic process for making polyurethanes was first developed in 1937 by the German chemist Otto Bayer (1902-1982), who patented his discovery and founded a company for its commercial production. The polyurethanes were first put to widespread use during World War II as substitutes for natural rubber in the manufacture of tires for military uses. The first rigid polyurethane foam was made in 1940, again for military purposes; the first polyurethane adhesive, for the joining of rubber and glass, in 1941; and the first use of polyurethane as an insulator, in beer barrels, in 1948. Polyurethane fibers were also used during World War II for the manufacture of protective clothing to be worn by soldiers in case of attacks by poison gas. In the late 1950s, a stretchable material made of polyurethane called spandex was introduced.

HOW IT IS MADE

Polyurethanes are formed by a reaction known as rearrangement. In this reaction, a hydrogen atom from the polyol (alcohol with two or more hydroxyl, -OH, groups) leaves the polyol and attaches itself to one of the nitrogen atoms in the diisocyanate molecule. The remnant of the polyol left after the hydrogen atom leaves then attaches itself to the carbon atom next to the nitrogen that has just received the hydrogen. The end result of this reaction is that two molecules, a diisocyanate molecule and a polyol molecule, have joined to make a single molecule. But that molecule has the same general structure as the original

Interesting Facts

The U.S. polyurethane industry produced 2,915 million kilograms (6,393 million pounds) of polyurethane in 2002.

reactants. So the reaction can be repeated with a second diisocyanate molecule and a second polyol molecule adding to the former product.

All that is needed to make this reaction happen is a suitable catalyst, a substance that will encourage the first hydrogen atom to leave the polyol and move to the diisocyanate. The catalyst most frequently used is diazobicyclo[2.2.2]octane, more commonly known as DABCO. DABCO provides an initial "tug" on the hydrogen atom that needs to be moved. Once it gets the reaction started, DABCO has no further role and can be recovered for re-use.

Various forms of polyurethane are made by adding additional steps to this fundamental process. For example, polyurethane foams are made by adding water and carbon dioxide to the molten polymer. The carbon dioxide creates bubbles, which makes the mixture rise like bread and then harden into a foam. Flexible polyurethane for fibers and foam rubber can be made by adding polyethylene glycol, a softening agent, to the mixture.

COMMON USES AND POTENTIAL HAZARDS

Flexible and rigid foams are the most popular types of polyurethane made in the United States. Flexible foams are used for cushioning, as in mattresses, upholstered furniture, and automobile seats. Semiflexible foams are component of carpet pads, sponges, packaging, and door panels for automobiles. Rigid polyurethane foams are used to produce insulation for roofs, refrigerators, and freezers. The construction industry accounted for more than half of all the rigid polyurethane foam made in 2002, primarily for

Words to Know

POLYMER a compound consisting of very large molecules made of one or two small repeated units called monomers.

THERMOPLASTIC capable of being repeatedly softened and hardened by alternately heating and cooling.

roofing and wall insulation. The automotive industry was the second largest user of rigid foams, where polyurethane was used for floor cushions, headliners, and heating and air-conditioning systems.

Polyurethane elastomers are flexible and strong. They can be stretched to a significant amount before returning to their original length. They are also shock absorbent. These properties account for their uses in tires, shoe soles, skateboards, and inline skates. Polyurethane elastomers can also be spun into fibers which are then used to make light, flexible clothing. Spandex, one of the most popular fabrics made of polyurethane fibers, is both sturdy and flexible. It is used to make bathing suits and exercise clothing because it can be distorted without losing its original shape or tearing. Polyurethane is also one of the main ingredients in the fabric known as pleather, or plastic leather. This synthetic form of leather is less expensive to produce and easier to dye than real leather.

Polyurethane coatings are strong and durable. They can be found on surfaces that need to be flexible, abrasion resistant, and chemical resistant, such as dance floors and bowling alleys. Water-based coatings are often used on airplanes and other transportation equipment. Polyurethane powder coatings are also used on fluorescent lights, refrigerators, car wheels and trim, lawnmowers, patio furniture, and ornamental iron.

Dust formed by the breakdown of polyurethane products may be an irritant to skin, eyes, and the respiratory system. No serious health problems have been associated with such materials, however. Some of the raw materials used in the manufacture of the polyurethanes, such as the isocyanates, do pose health risks. These risks are of concern primarily to workers who come into contact with those raw materials.

FOR FURTHER INFORMATION

Alliance for the Polyurethane Industry. "About Polyurethane" http://www.polyurethane.org/about/(accessed on October 26, 2005).

"Making Polyurethanes." Polymer Science Learning Center, University of Southern Mississippi. http://www.pslc.ws/mactest/uresyn.htm (accessed on October 26, 2005).

"Polyurethanes." Huntsman. http://www.huntsman.com/pu/ (accessed on October 26, 2005).

Vartan, Starre. "Pretty in Plastic: Pleather Is a Versatile, though Controversial, Alternative to Leather." *E* (September-October 2002): 53-54.

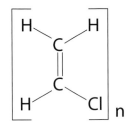

OTHER NAMES:
See Overview.

FORMULA:
-[-CH$_2$CHCl-]-$_n$

ELEMENTS:
Carbon, hydrogen, chlorine

COMPOUND TYPE:
Organic polymer

STATE:
Solid

MOLECULAR WEIGHT:
Varies

MELTING POINT:
Decomposes at 148°C (298°F)

BOILING POINT:
Not applicable

SOLUBILITY:
Insoluble in water; soluble in tetrahydrofuran, dimethyllformamide, dimethylsulfoxide

KEY FACTS

Polyvinyl Chloride

OVERVIEW

Polyvinyl chloride (pol-ee-VYE-nul KLOR-ide) is also known as PVC, vinyl, chlorethylene homopolymer, and chlorethene homopolymer. It is the third most commonly produced plastic in the United States, exceeded only by polyethylene and polypropylene. It is offered commercially in a variety of formulations, usually as a white powder or colorless granules. The compound is resistant to moisture, weathering, most acids, fats and oils, many organic solvents, and attack by fungi. It is easily colored and manufactured in a variety of forms, including sheets, films, fibers, and foam.

Polyvinyl chloride was first discovered accidentally in 1835 by the French physicist and chemist Henry Victor Regnault (1810-1878). Regnault found that a container of gaseous vinyl chloride (CH$_2$CH=Cl) exposed to the sunlight gradually changed to a white powder. Regnault knew almost nothing about the composition of the powder or how it was formed. Polyvinyl chloride remained a subject of little or no interest to chemists for almost a century. German and

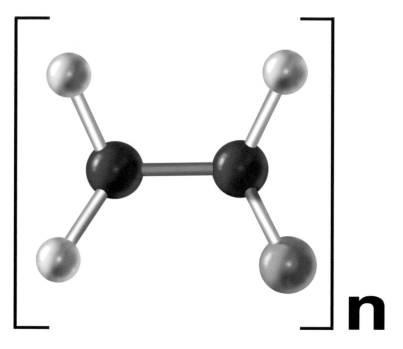

Poly(vinyl chloride). White atoms are hydrogen, black atoms are carbon and green atom is chlorine. Gray stick represents a double bond.
PUBLISHERS RESOURCE GROUP

Russian chemists made some efforts to find useful applications for the compound in the early twentieth century, without much success. The first patent for the production of the compound was awarded to the German chemist Friedrich Heinrich August Klatte (dates not available) in 1913, but Klatte never marketed the product for commercial use.

Then, in 1926, Waldo Lonsbury Semon (1898-1999), a chemist at B. F. Goodrich, rediscovered polyvinyl chloride, once again by accident. Semon had been assigned the task at Goodrich of finding a substitute for natural rubber as a lining for metal tanks and for finding a material that would bond the rubber substitute to the metal tank. One of the materials he studied was polyvinyl chloride. In his work, Semon did not so much rediscover polyvinyl chloride as to find new ways of working with the material. For example, he found that he could dissolve the compound in various organic solvents, converting them so that they could be molded, extruded, and formed. He also discovered that he could control the properties of the material by altering the amount of solvent used to dissolve powdered polyvinyl chloride.

Interesting Facts

- The recycling symbol for polyvinyl chloride is the number 3 inside a triangle made of three arrows.

- The environmental group Greenpeace has called for a global ban on the production of polyvinyl chloride because the toxic substance dioxin is released (albeit, in very small amounts) as a byproduct of the compound's production.

Goodrich adapted Semon's discovery for two specific applications: shoe heels and the coating on chemical racks. Those applications were not profitable enough for Goodrich to continue making polyvinyl chloride. But Semon continued to look for new ways of adapting the compound for additional applications. He was eventually successful and by 1931 the company had begun to turn out a full line of polyvinyl chloride products in most of the forms currently available.

HOW IT IS MADE

Polyvinyl chloride is made by polymerizing vinyl chloride (CH_2=CHCl). Polymerization occurs when the double bond in vinyl chloride breaks, allowing one molecule of vinyl chloride to combine with a second molecule of vinyl chloride: CH_2=CHCl + CH_2=CHCl → CH_3CHClCH=CHCl. The product of that reaction also contains a double bond, allowing the reaction to be repeated: CH_3CHClCH=CHCl + CH_2=CHCl → CH_3CHClCH$_2$CHClCH=CHCl. Once again, the final product contains a double bond, and the reaction can be repeated again and again and again. The reaction is made possible by using some agent to cause double bonds to break. In the original experiments carried out by Regnault, Klatte, and others, that agent was sunlight. Chemists have long since learned, however, that a variety of chemicals known as peroxide initiators are more effective at breaking double bonds. Peroxide initiators are compounds with an oxygen-oxygen bond (-O-O-). One of the most widely used initiators is benzoyl peroxide (C_6H_5CO-O-O-C_6H_5CO).

COMMON USES AND POTENTIAL HAZARDS

An estimated 7 billion kilograms (15.7 billion pounds) of polyvinyl chloride were produced in the United States in 2006. It is available in more than 50 trade names, including Airex, Armodour, Astralon, Benvic, Bonloid, Chemosol, Chlorostop, Dacovin, Dorlyl, Flocor, Lucoflex, Norvinyl, Opalon, Polivinit, Polytherm, Sicron, Takilon, Vinikulon, Viniplast, and Wilt Pruf. About three-quarters of that amount was made in a rigid format that is hard and inflexible. The remaining one-quarter was made in a flexible form, produced by adding materials to polyvinyl chloride that make it soft and pliable. About 75 percent of all rigid polyvinyl chloride (half of all the compound made in the United States) goes to the construction industry. It has replaced older building materials such as clay, concrete, and wood because it is inexpensive, lightweight, resistant to damage by the sun, and easy to assemble. The compound is used to make vinyl siding, windows, plumbing pipes, flooring, electric cables, roofing materials, and insulation for cables and wires.

Flexible polyvinyl chloride is used to make fibers and films for applications such as clothing, upholstery, plastic bottles, medical equipment, lightweight toys, shower curtains, and packaging films. Some of the medical equipment produced from polyvinyl chloride include bags to hold blood and other fluids, artificial heart valves, and tubes used in kidney dialysis. Medical products made from polyvinyl chloride are strong enough to be air-dropped to troops in combat zones. Automobile manufacturers use polyvinyl chloride in body side moldings, interior upholstery, engine wiring, floor mats, adhesives, dashboards, arm rests, and coatings under the vehicle.

The commercial and household products containing polyvinyl chloride are generally regarded as posing no threat to human health. However, a number of questions have been raised about possible health hazards and risks to the environment as a result of the process by which polyvinyl chloride is made, some of its applications, and its eventually disposal. For example, polyvinyl chloride is made from vinyl chloride, which itself is toxic and a carcinogen. People who work with vinyl chloride in production facilities are at risk for developing a form of liver cancer that may be related to exposure to vinyl chloride. Vinyl chloride, in turn, is made from chloride, a very toxic gas that poses health risks to people who work with it.

Some environmental health experts point out that products made of polyvinyl chloride may give off toxic or carcinogenic gases for short periods of time after they have been put into use. For example, shower curtains and automobile upholstery containing polyvinyl chlorides may release hazardous chemicals into the air for a few months after they are first installed.

Some concern has also been expressed about additives used with polyvinyl chloride. For example, some of the substances used to soften the compound belong to a family known as the phthalates, which are known carcinogens with toxic effects. Some health experts point out that infants may ingest small amounts of phthalates from PVC toys they chew on.

Finally, the disposal and destruction of compounds containing polyvinyl chloride may create environmental problems. Burning such products, for example, releases hydrogen chloride gas, a suffocating and toxic gas, into the atmosphere. Enough concern about the health and environmental hazards has arisen that some governmental bodies in Europe have placed limitations on the uses to which PVC products can be put.

FOR FURTHER INFORMATION

"Healthy Building Network."
http://www.healthybuilding.net/ (accessed on October 29, 2005).

Meikle, Jeffrey L. *American Plastic: A Cultural History*. Piscataway, N.J.: Rutgers University Press, 1997.

Thornton, Joe. "Environmental Impacts of Polyvinyl Chloride (PVC) Building Materials." A briefing paper for the U.S. Green Building Council, n.d. Available online at http://www.usgbc.org/Docs/LEED_tsac/PVC/CMPBS%20Original%20Submittal.pdf (accessed on October 29, 2005).

"Vinyl—The Material." The Vinyl Institute.
http://www.vinylinfo.org/materialvinyl/history.html (accessed on October 29, 2005).

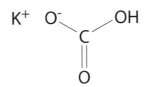

Potassium Bicarbonate

OVERVIEW

Potassium bicarbonate (poe-TAS-ee-yum buy-KAR-bo-nate) is a colorless crystalline solid or white powder with no odor and a salty taste. It occurs naturally in salt beds, sea water, silicate rocks, and a number of foods, primarily fruits and vegetables. Potassium bicarbonate is also present in the tissues of humans and other animals, where it is involved in a number of essential biological processes, including digestion, muscle contraction, and heartbeat. It is used primarily in cooking and baking, as a food additive, and in fire extinguishers.

HOW IT IS MADE

Potassium bicarbonate is made by passing carbon dioxide gas through an aqueous solution of potassium carbonate:

$$K_2CO_3 + CO_2 + H_2O \rightarrow 2KHCO_3$$

COMMON USES AND POTENTIAL HAZARDS

One of the most familiar applications of potassium bicarbonate is as an antacid to treat the symptoms of upset stomach. The compound reacts with stomach acid–hydrochloric acid;

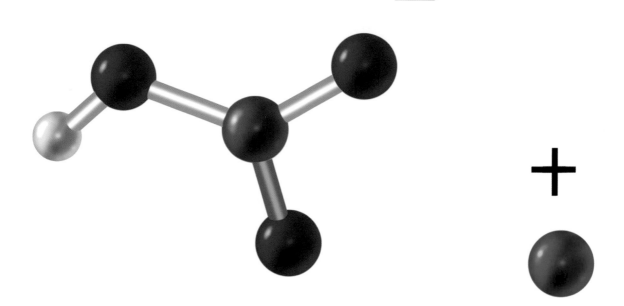

Potassium bicarbonate. Red atoms are oxygen, white atom is hydrogen; black atom is carbon; and turquoise atom is potassium. Gray stick indicates double bond. PUBLISHERS RESOURCE GROUP

HCl—to relieve gaseous distress, stomach pain, and heartburn. The compound can also be used to treat potassium deficiency in the body. Some research suggests that potassium bicarbonate may help restore muscle and bone tissue, particularly in women with the degenerative bone disease osteoporosis. The compound is also used as a food additive, as a leavening agent, to maintain proper acidity in foods, to supply potassium to a diet, and to provide the bubble and fizz in carbonated drinks.

Potassium bicarbonate is also used in certain types of fire extinguishers. When such an extinguisher is used, the potassium bicarbonate reacts with an acid present in the device to produce carbon dioxide. The carbon dioxide propels a liquid from the extinguisher and, itself, helps put out a fire. Potassium bicarbonate is also used in agriculture to maintain proper acidity in soils and to supply potassium that may be missing from the ground.

Under normal circumstances, potassium bicarbonate poses no health threat to humans. Excess potassium in the body may result in a condition known as hyperkalemia, characterized by tingling of the hands and feet, muscle weakness, and temporary paralysis. Such a condition is very rare when potassium bicarbonate is used in normal amounts.

Interesting Facts

Potassium bicarbonate can be substituted for baking soda (sodium bicarbonate; NaHCO$_3$) for people who are on a low-sodium diet.

FOR FURTHER INFORMATION

"Potassium Bicarbonate." Yale New Haven Health Drug Guide. http://yalenewhavenhealth.org/library/healthguide/en-us/drugguide/topic.asp?hwid=d036ooa1 (accessed on October 31, 2005).

"Potassium Bicarbonate Handbook." Armand Products Company. Available online at http://www.oxy.com/OXYCHEM/Products/potassium_bicarbonates/literature/PootBiVs6.pdf (accessed on October 31, 2005).

Rowley, Brian. "Fizzle or Sizzle? Potassium Bicarbonate Could Help Spare Muscle and Bone." *Muscle & Fitness* (December 2002): 72.

"Strong Muscle and Bones." *Prevention* (June 1, 1995): 70-73.

See Also Sodium Bicarbonate

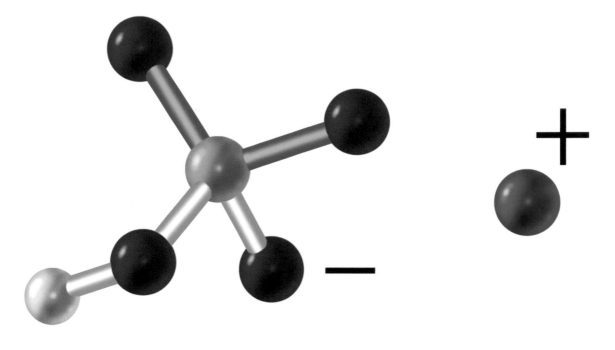

Potassium bisulfate. Red atoms are oxygen; white atom is hydrogen; yellow atom is sulfur; and turquoise atom is potassium. Gray sticks indicate double bonds. PUBLISHERS RESOURCE GROUP

list of Generally Regarded as Safe (GRAS) list. The list contains chemicals thought to be safe for human consumption even though they have not been tested. Potassium bisulfate is used in foods as a preservative because it interferes with the growth of insects, bacteria, and fungi that cause foods to spoil. It is also used as a leavening agent in cake mixes. One of its most important uses is in the wine industry, where it is used to convert certain compounds that occur naturally in grapes into potassium bitartrate. Potassium bisulfate is also used as a flux, in the analysis of ores and silica compounds, in the manufacture of fertilizers, and in the preparation of methyl and ethyl acetate.

Potassium bisulfate is a strong irritant to human tissue. If spilled on the skin, inhaled, or ingested, it can burn tissue causing skin rashes, sore nasal passages, irritation of the throat, and damage to the eyes. Burns of the mouth and stomach may also occur. These hazards are of concern primarily to people who work directly with the compound and do not pose a threat as a food additive.

CHEMICAL COMPOUNDS

Words to Know

DELIQUESCENT having a strong tendency to absorb moisture from the air, so that it becomes wet and dissolves in the water it has absorbed.

FLUX a material that lowers the melting point of another substance or mixture of substances or that is used in cleaning a metal.

FOR FURTHER INFORMATION

"Agency Response Letter: GRAS Notice No. GRN 000060." U.S. Food and Drug Administration. http://www.cfsan.fda.gov/~rdb/opa-g060.html (accessed on November 1, 2005).

"Safety (MSDS) data for potassium bisulfate ." Physical and Theoretical Chemistry Laboratory, University of Oxford. http://ptcl.chem.ox.ac.uk/MSDS/PO/potassium_bisulfate.html (accessed on November 1, 2005).

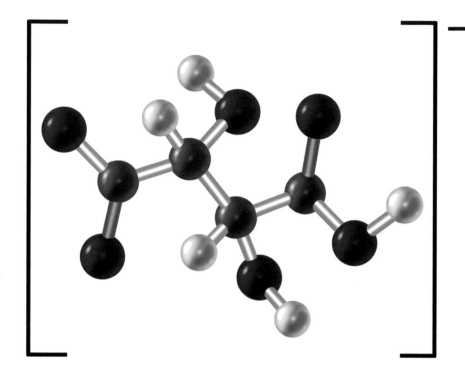

Potassium bitartrate. Red atoms are oxygen; white atoms are hydrogen; black atoms are carbon; and blue atom is potassium. Gray sticks indicate double bonds. PUBLISHERS RESOURCE GROUP

after grapes have been crushed to make wine) are treated with hot water, which dissolves the potassium bitartrate. The hot solution is then allowed to evaporate. As potassium bitartrate crystals form, they are removed and purified.

COMMON USES AND POTENTIAL HAZARDS

The primary use of potassium bitartrate is in commercial food production and household cooking and baking. The compound is added to foods to stabilize egg whites after they have been beaten, to add body to a product, and to produce creamier textures for sugar-based foods. A small amount of potassium bitartrate also adds a pleasantly acidic taste to foods. The compound is found most commonly in baked goods, candies, crackers, confections, gelatins, puddings, jams and jellies, soft drinks, margarines, and frostings. Some of the functions that potassium bitartrate performs in foods include:

- It acts as a leavening agent, causing a product to rise, in baked goods;

- It serves as an anticaking agent and stabilizer to thicken some food products;

- It prevents the crystallization of sugars when making candy and sugary syrups;

- It hides the harsh aftertaste and intensifies the flavors of some foods; and

- It improves the color of vegetables that have been boiled during preparation.

Potassium bitartrate also has a number of other uses. It can be used as a laxative for both humans and domestic animals. It is used in the processing of some metals, giving a colored tinge to the final product. The compound is also an ingredient in products used to clean brass, copper, aluminum, and other metals. And it is used in the chemical industry as a raw material for the preparation of other tartrate compounds.

There are no known health hazards for potassium bitartrate except for the general overall caution given in the Introduction that all chemicals in some concentrations can pose a hazard. Potassium bitartrate is thought to be one of the safest chemical compounds for general use.

FOR FURTHER INFORMATION

"Potassium Bitartrate." ChemicalLand21.com. http://www.chemicalland21.com/arokorhi/lifescience/foco/POTASSIUM%20BITARTRATE.htm (accessed on October 29, 2005).

"Potassium Bitartrate." J. T. Baker. http://www.jtbaker.com/msds/englishhtml/p5565.htm (accessed on October 29, 2005).

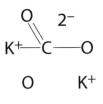

Potassium Carbonate

OVERVIEW

Potassium carbonate (poe-TAS-ee-yum KAR-bun-ate) is also known as potash, pearl ash, salt of tartar, carbonate of potash, and salt of wormwood. It is a white, translucent, odorless, granular powder or crystalline material that tends to absorb water from the air. As it does, it is converted into the sesquihydrate ("sesqui" = one-and-a-half) with the formula $K_2CO_3 \cdot 1.5H_2O$. That formula means that three molecules of potassium carbonate share two molecules of water among them.

Potash is easily produced by pouring water over the ashes of burned plants and then evaporating the solution formed in large pots (hence the name: "pot" "ash"). The process has been known since at least the sixth century CE and the resulting product used in the manufacture of soap. Potash was one of the first chemicals to be exported by American colonists, with shipments having left Jamestown as early as 1608. Potassium carbonate is also called pearl ash and salt of tartar, both of which are impure forms of the compound. The impurities present include sodium chloride

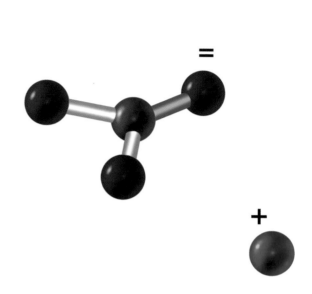

Potassium carbonate. Red atoms are oxygen; black atom is carbon; and turquoise atoms are potassium. Gray stick indicates a double bond.
PUBLISHERS RESOURCE GROUP

and some heavy metals (such as iron and lead). The primary uses of potassium carbonate are in the production of fertilizers, soaps, and heat-resistant glass.

HOW IT IS MADE

Most of the potassium carbonate made in the United States is produced beginning with potassium chloride (KCl) obtained from seven mines in New Mexico, Michigan, and Utah. The potassium chloride is first converted to potassium hydroxide (KOH) by electrolysis. The potassium hydroxide is then treated with carbon dioxide (CO_2) to obtain potassium bicarbonate ($KHCO_3$). Finally, the potassium bicarbonate is decomposed by heating, yielding water, carbon dioxide, and potassium carbonate.

Another method of preparation, called the Engel-Precht process, is a modification of this procedure. A mixture of potassium chloride, magnesium carbonate or magnesium oxide, and carbon dioxide is treated under 30 atmospheres

Interesting Facts

- The first patent ever issued in the United States was awarded in 1790 to Samuel Hopkins for a new and better way of making pearl ash.

- Pearl ash was used in the United States in the eighteenth century as a leavening agent in the baking of bread.

- The demand for potash began to fall off in the late eighteenth century as improved methods for the synthesis of sodium carbonate were developed. Sodium carbonate can replace potassium carbonate in many applications.

- The term potash has historically had many different meanings. It has been used to refer to potassium hydroxide (KOH), potassium chloride (KCl), potassium sulfate (K_2SO_4), potassium nitrate (KNO_3), or to some combination of these compounds.

of pressure, with the formation of a double salt, $KHCO_3 \cdot MgCO_3 \cdot 4H_2O$. The double salt is then heated to obtain potassium carbonate. The traditional method of obtaining potash from wood and vegetable ash is now obsolete.

COMMON USES AND POTENTIAL HAZARDS

An estimated 7 million metric tons (6.5 million short tons) of potassium carbonate were produced in the United States in 2005. Of that amount, nearly 90 percent was used for the production of fertilizers. Potash provides plants with the potassium they need to stay healthy and grow. Potassium is one of the three major nutrients required by plants, the other two being nitrogen and phosphorus. The next largest application for potassium carbonate is in the chemical industry, where it is used as a raw material to make other chemical compounds, potassium silicate being the most common.

Smaller amounts of potassium carbonate are still used for what was once its major application: the manufacture of soap. Potassium soaps (made from potassium carbonate) have

Words to Know

CAUSTIC Strongly basic or alkaline; can irritate or corrode living tissue.

ELECTROLYSIS Process in which an electric current is used to bring about chemical changes.

FLUX A material that lowers the melting point of another substance or mixture of substances or that is used in cleaning a metal.

some characteristics different from more common sodium soaps (made from sodium carbonate). They tend to be softer or even liquid and better able to create suds in water that contains a high concentration of minerals. Potassium carbonate is also used to make specialty glasses, such as television screens, cathode ray tubes, and optical lenses. Some other uses of the compound include:

- For glazes in the making of pottery;
- In the manufacture of pigments and printing inks;
- As an additive in certain food products, chocolate being one example;
- For the tanning and finishing of leather and the dyeing, washing, and finishing of wool; and
- As a flux in metal working.

Potassium carbonate in dry or dissolved form is an irritant to the eyes, skin, and respiratory system. It can cause inflammation of the skin, eyes, throat, and stomach. Potassium carbonate's action is caused by its caustic properties in water solution, which are produced when it is dissolved in water or when it is absorbed by moist tissues in the body.

FOR FURTHER INFORMATION

"International Potash Institute."
http://www.ipipotash.org/index.php (accessed on October 31, 2005).

"The Potash Trade." Townships Heritage WebMagazine.
http://www.townshipsheritage.com/Eng/Hist/Life/potash.html (accessed on October 31, 2005).

"Potassium Carbonate." Chemical Land 21. http://www.chemicalland21.com/arokorhi/industrialchem/inorganic/POTASSIUM%20CARBONATE.htm(accessed on October 31, 2005).

"Potassium Carbonate." J. T. Baker. http://www.jtbaker.com/msds/englishhtml/p5609.htm (accessed on October 31, 2005).

Willett, Jason C. "Potash." U.S. Geological Survey Commodity Statistics and Information. Available online at http://minerals.usgs.gov/minerals/pubs/commodity/potash/potasmyb04.pdf (accessed on October 31, 2005).

See Also Potassium Chloride; Potassium Hydroxide

Cl⁻   K⁺

Potassium Chloride

OVERVIEW

Potassium chloride (poe-TAS-ee-yum KLOR-ide) occurs as a white or colorless crystalline solid or powder. It is odorless, but has a strong saline (salty) taste. It occurs naturally in the minerals sylvite, carnallite, kainite, and sylvinite. It also occurs in sea water at a concentration of about 0.076 percent (grams per milliliter of solution). Potassium chloride is the most abundant compound of the element potassium and has the greatest number of applications of any salt of potassium. By far the most important application of potassium chloride is in the manufacture of fertilizers.

HOW IT IS MADE

All of the major sources of potassium chloride have their origin in sea water. Sea water is a solution of a number of salts dissolved in water. The most important of those salts are sodium chloride (about 2.3 percent), magnesium chloride (about 0.5 percent), sodium sulfate (about 0.4 percent),

Potassium chloride. Turquoise atom is potassium and green atom is chlorine. Potassium atom is positively charged. Chlorine atom is negatively charged. PUBLISHERS RESOURCE GROUP

calcium chloride (about 0.1 percent) and potassium chloride (about 0.07 percent). When large bodies of sea water dry up, they leave behind complex mixtures of minerals consisting of these salts. Over millions of years, huge deposits of these minerals have been buried under the land. In the United States, sea salt deposits are found in New Mexico, Texas, California, and Michigan.

Any one of the salts present in a sea salt deposit—including potassium chloride—can be extracted by a common procedure. The minerals that make up the deposit are crushed and dissolved in hot water. The solution is then allowed to cool very slowly. As it cools, each of the dissolved salts crystallizes out at a specific temperature, is removed from the solution, and is purified. Since potassium chloride is much more soluble in hot water than in cold water, it crystallizes out after other salts have been removed.

The majority of potassium chloride in the United States is now extracted by a lengthy process that also begins with the crushing of natural ores, such as sylvite and carnalite. The solid mixture is then cleaned and purified before being treated with a flotation agent, usually some type of amine. A flotation agent is a material that coats the desired compound, such as potassium chloride, and allows it to float to the surface of the reaction chamber, like the soap suds that float on top of a washing machine. An amine is an organic compound that contains the nitrogen, usually as the -NH_2

Interesting Facts

- One use of potassium chloride is as a lethal injection for prisoners who have been given the death penalty. The chemical interferes with normal heart function and causes a heart attack within five to about eighteen minutes after injection. Thirty-four of the United States prescribe death by lethal injection for prisoners who have been convicted of murder.

group. The amine-coated potassium chloride is skimmed off the top of the reaction mixture, purified, and prepared in some crystalline or powder form.

COMMON USES AND POTENTIAL HAZARDS

Potassium chloride is present in some foods in small amounts. The compound is also used as a food additive to increase the acidity and to stabilize, thicken, or soften some food products, such as jams and jellies and preserves that are artificially sweetened. Many infant formulas also contain potassium chloride. Potassium chloride is also used as a nutrient for yeast cultures and in making beer. The compound is used as a salt substitute for people who are on low-salt (meaning low-sodium) diets. Some brand names of these products are LoSalt®, Reheis Less Salt Blend®, and Morton® Lite Salt®.

The largest application of potassium chloride is in the production of fertilizers. More than ninety percent of the potassium chloride produced in the United States is used for that purpose. The compound provides the potassium plants need to stay healthy and grow normally. It is one of three macronutrients—substances needed in relatively large amounts—for normal growth. The other two macronutrients are phosphorus and nitrogen. Smaller amounts of potassium chloride are used in the production of other potassium compounds, in photography, and in chemical research applications.

FOR FURTHER INFORMATION

Benfell, Carol. "Routine but Deadly Drug: Potassium Chloride Has a Jekyll and Hyde Personality." American Iatrogenic Association. http://www.iatrogenic.org/potchlor.html (accessed on October 31, 2005).

"Potassium Chloride." MedicineNet.com. http://www.medicinenet.com/potassium_chloride/article.htm (accessed on October 31, 2005).

"Potassium Chloride." University of Maryland Medical Center. http://www.umm.edu/altmed/ConsDrugs/PotassiumChloridecd.html (accessed on October 31, 2005).

See Also Potassium Carbonate; Potassium Hydroxide; Potassium Nitrate; Sodium Chloride

$$K^+ \qquad F^-$$

KEY FACTS

OTHER NAMES:
Potassium mono-
fluoride

FORMULA:
KF

ELEMENTS:
Potassium, fluorine

COMPOUND TYPE:
Binary salt (inor-
ganic)

STATE:
Solid

MOLECULAR WEIGHT:
58.10 g/mol

MELTING POINT:
$858°C (1580°F)$

BOILING POINT:
$1502°C (2736°F)$

SOLUBILITY:
Soluble in cold water;
very soluble in hot
water; insoluble in
ethyl alcohol; soluble
in hydrofluoric acid
(H_2F_2)

Potassium Fluoride

OVERVIEW

Potassium fluoride (poe-TAS-ee-yum FLU-ride) is a color-less or white crystalline or powdery compound with no odor, but a sharp, salty taste. It has somewhat limited uses in industry and chemical research.

HOW IT IS MADE

In one method for making potassium fluoride, potassium carbonate (K_2CO_3) is dissolved in hydrofluoric acid, resulting in the formation of potassium bifluoride (KHF_2): $K_2CO_3 + 2H_2F_2 \rightarrow 2KHF_2 + CO_2 + H_2O$. The potassium bifluoride is then heated to form potassium fluoride and hydrogen fluoride: $KHF_2 \rightarrow KF + HF$.

Potassium fluoride can also be prepared by the direct reaction between hydrofluoric acid and potassium hydroxide: $H_2F_2 + 2KOH \rightarrow 2KF + 2H_2O$. The potassium fluoride thus formed is then dried and crystallized or converted to powder form.

Potassium fluoride. Silver atom is potassium; yellow atom is fluorine. PUBLISHERS RESOURCE GROUP

COMMON USES AND POTENTIAL HAZARDS

Potassium fluoride is used as a fluoridating agent—a substance that provides fluorine atoms to other compounds—in the preparation of organic chemicals. It also finds some use in the field of metallurgy, where it is used as a flux, to finish metals, to make coatings for metals, and in tin

Interesting Facts

• In France, potassium fluoride is sometimes added to table salt to help prevent dental cavities.

• Potassium fluoride can not be shipped out of the United States to other countries without a special license from the U.S. Department of Commerce because the compound is an important raw material in the manufacture of certain chemical weapons.

Words to Know

FLUX A material that lowers the melting point of another substance or mixture of substances or that is used in cleaning a metal.

METALLURGY The study of the properties and structures of metals.

plating. Potassium fluoride is used to frost and etch glass, as in the manufacture of some optical glasses, and to make insecticides, pesticides, and disinfectants.

Potassium fluoride is irritating to the skin, eyes, and respiratory system. It is moderately toxic by ingestion, causing nausea, vomiting, diarrhea, and stomach pains. In larger doses, it can cause damage to the brain, kidneys, and heart. Long-term exposure to potassium fluoride can cause damage to the teeth and bones. One condition that can develop is called fluorosis. Symptoms of the condition include brittle bones, weight loss, anemia, hardening of the ligaments, and stiffness of the joints.

FOR FURTHER INFORMATION

Hernandez, Lucúla Pazos. "Prevention of Dental Caries Through Salt Fluoridation in Mexico." http://www.ibiblio.org/taft/cedros/english/newsletter/n3/prevent.html (accessed on October 31, 2005).

"Potassium Fluoride." Patnaik, Pradyot. *Handbook of Inorganic Chemicals.* New York: McGraw-Hill, 2003, 754-755.

"Potassium Fluoride, Anhydrous." J. T. Baker. http://www.jtbaker.com/msds/englishhtml/p5774.htm (accessed on October 31, 2005).

"Silver Production on the Moon." The Artemis Project. http://asi.org/adb/02/13/02/silicon-production.html (accessed on October 31, 2005).

K$^+$

$$O\!-\!\!\!\underset{}{}\text{H}^-$$

OTHER NAMES:
Caustic potash;
potash lye; potassa;
potassium hydrate

FORMULA:
KOH

ELEMENTS:
Potassium, oxygen,
hydrogen

COMPOUND TYPE:
Base (inorganic)

STATE:
Solid

MOLECULAR WEIGHT:
56.10 g/mol

MELTING POINT:
406°C (763°F)

BOILING POINT:
1327°C (2421°F)

SOLUBILITY:
Soluble in water,
ethyl alcohol, methyl
alcohol, and glycerol

KEY FACTS

Potassium Hydroxide

OVERVIEW

Potassium hydroxide (poe-TAS-ee-yum hy-DROK-side) is a white deliquescent solid that is available in sticks, lumps, flakes, or pellets. A deliquescent material is one that tends to absorb so much moisture from the atmosphere that it becomes very wet, even to the point of dissolving in the water it has absorbed. Potassium hydroxide also absorbs carbon dioxide from the air, changing in the process to potassium carbonate (K_2CO_3). Potassium hydroxide is one of the most caustic materials known. It has a number of uses in industry and agriculture.

Potassium hydroxide is chemically very active. It reacts violently with acids, generating significant amounts of heat in the process. In moist air, it corrodes metals such as tin, lead, zinc, and aluminum with the release of combustible and explosive hydrogen gas.

HOW IT IS MADE

Potassium hydroxide is made by the electrolysis of an aqueous solution of potassium chloride (KCl). In that process,

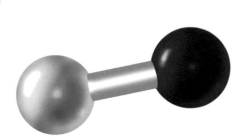

Potassium hydroxide. Red atom is oxygen; white atom is hydrogen; and turquoise atom is potassium. PUBLISHERS RESOURCE GROUP

an electric current decomposes potassium chloride into potassium and chlorine. The chlorine escapes as a gaseous by-product and the potassium reacts with water to form potassium hydroxide.

COMMON USES AND POTENTIAL HAZARDS

An estimate 440,000 metric tons (485,000 short tons) of potassium hydroxide were used in the United States in 2005. About 53 percent of that amount was used in the production of other potassium compounds, especially potassium carbonate (28 percent), potassium acetate, potassium cyanide, potassium permanganate, and potassium citrate. About 10 percent of all caustic potash was used in the manufacture of potassium soaps and detergents. Most soaps and detergents are made of sodium hydroxide. But potassium hydroxide can be substituted for sodium hydroxide to obtain soaps and detergents with special properties. Liquid soaps and soaps that will lather in salt water or water with a high mineral content are examples of such specialized potassium soaps.

Some other applications of potassium hydroxide include:

- The manufacture of liquid fertilizers, herbicides, and other agricultural chemicals;

- As a neutralizing agent in many chemical and industrial processes;

- In the production of synthetic rubber;

Interesting Facts

- Pure potassium hydroxide is difficult to prepare since the compound is so reactive that it tends to react with moisture, carbon dioxide, and other impurities with which it comes into contact. The compound is commercially available in a purity of about 90 percent. Much purer products are available, however, when needed.

- In the food production industry, where it is used in the removal of peels from fruits and vegetables, the carmelization of products that contain sugar, the thickening of ice cream, the softening of olives, the production of chocolate and cocoa, and the preparation of hominy from corn kernels;

- In the manufacture of alkaline storage batteries and some types of fuel cells;

- In the refining of petroleum;

- As a way for removing horn buds from young cattle;

- In a number of cosmetic procedures, such as softening of cuticles, removal of warts, and cleaning of dentures; and

- As an ingredient in paint removers.

Potassium hydroxide is a very hazardous chemical. It is corrosive to tissue and can cause severe burns of the skin, eyes, and mucous membranes. If ingested, it can cause internal bleeding, scarring of tissue, nausea, vomiting, diarrhea, and lowered blood pressure that can result in a person's collapse. In sufficient amounts, it can cause death. Inhalation of potassium hydroxide fumes or dust can cause lung irritation, sneezing, sore throat, runny nose, and severe damage to the lungs. In contact with the eyes, the compound can cause blurred vision and, in sufficient amounts, loss of eyesight. People who have to work with the compound should always wear goggles, gloves, and protective clothing to reduce their risk of contact with the chemical.

Words to Know

AQUEOUS SOLUTION A solution that consists of some material dissolved in water.

CAUSTIC Strongly basic or alkaline; able to irritate or corrode living tissue.

DELIQUESCENT Having the tendency to absorb moisture and, therefore, dissolve or melt.

MUCOUS MEMBRANES Tissues that line the moist inner lining of the digestive, respiratory, urinary and reproductive systems.

FOR FURTHER INFORMATION

"Caustic Potash." Occidental Petroleum Corporation. http://www.oxy.com/OXYCHEM/Products/caustic_potash/caustic_potash.htm (accessed on November 1, 2005).

Cavitch, Susan Miller. The Natural Soap Book: Making Herbal and Vegetable-Based Soaps. Markahm, Canada: Storey Publishing, 1995.

"Potassium Hydroxide." Medline Plus. http://www.nlm.nih.gov/medlineplus/ency/article/002482.htm (accessed on November 1, 2005).

"Potassium Hydroxide." NIOSH Pocket Guide to Chemical Hazards. http://www.cdc.gov/niosh/npg/npgd0523.html (accessed on November 1, 2005).

See Also Sodium Hydroxide

I^- K^+

Potassium Iodide

FORMULA:
KI

ELEMENTS:
Potassium, iodine

COMPOUND TYPE:
Binary salt (inor-
ganic)

STATE:
Solid

MOLECULAR WEIGHT:
166.00 g/mol

MELTING POINT:
681°C (1260°F)

BOILING POINT:
1323°C (2413°F)

SOLUBILITY:
Soluble in water,
ethyl alcohol, acet-
one, and glycerol

KEY FACTS

OVERVIEW

Potassium iodide (poe-TAS-ee-yum EYE-oh-dide) is a white crystalline, granular, or powdered solid with a strong, bitter, salty taste. It is used as a feed additive, a dietary supplement, in photographic films, and in chemical research.

HOW IT IS MADE

A number of methods are available for the preparation of potassium iodide. In one procedure, elemental iodine (I_2) is added to a solution of potassium hydroxide (KOH): $I_2 + 6KOH \rightarrow 5KI + KIO_3 + 3H_2O$. The potassium iodide formed is separated from the potassium iodate (KIO_3) by fractional crystallization. That is, the solution is warmed and then cooled. As the temperature falls, the two compounds, potassium iodide and potassium iodate, crystalize out at different temperatures and can be separated from each other. The potassium iodate can then be heated, causing it to decompose and make additional potassium iodide: $2KIO_3 \rightarrow 2KI + 3O_2$.

Potassium iodide. Green atom is iodine and turquoise atom is potassium. PUBLISHERS RESOURCE GROUP

Potassium iodide can also be produced by reacting hydriodic acid (HI) with potassium bicarbonate: HI + KHCO$_3$ → KI + CO$_2$ + H$_2$O.

Finally, the compound can be made by reacting iron (III) iodide with potassium carbonate (K$_2$CO$_3$): Fe$_3$I$_8$ + 4K$_2$CO$_3$ → 8KI + 4CO$_2$ + Fe$_3$O$_4$.

COMMON USES AND POTENTIAL HAZARDS

Potassium iodide is added to animal feeds to ensure that domestic animals get the iodine they need in their daily diets. A mixture of iodine in aqueous potassium iodide called SSKI has long been used as a disinfectant. The active agent in this mixture is iodine, and the potassium iodide is added to increase the iodine's solubility in water. SSKI is used to purify small amounts of water, to clear up pimples, to prevent sinus infections, and to treat bladder, lung, and stomach infections.

Perhaps the best known use of potassium iodide today is as a treatment for radiation exposure. When a nuclear bomb explodes or a nuclear accident occurs, one of the most dangerous products released to the environment is a radioactive isotope known as iodine-131. Iodine-131 enters the human body and travels to the thyroid, where it attacks cells and tissues, eventually resulting in thyroid cancer. Experts recommend that people exposed to radiation take potassium iodide as a protection against this hazard. The potassium iodide saturates

Interesting Facts

- Iodized salt contains a small amount of potassium iodide. The potassium iodide is added to ensure that people get enough iodine in their daily diet. A deficiency of iodine results in a medical condition known as goiter, a large swelling in the neck.

- Before iodized salt was available, many people failed to get enough iodine in their diets and goiter was a very common health problem.

- Potassium iodide occurs naturally in seaweed.

the thyroid gland, making it difficult for the gland to absorb radioactive iodine that may enter the body.

Potassium iodide is also used in the production of photographic film and in a number of chemical tests, such as the determination of dissolved oxygen in water and the presence of starch in an unknown mixture.

Under most circumstances, there are no health hazards associated with potassium iodide. Taking an excess of the compound may have harmful effects on the thyroid gland, however. For that reason, people with an overactive thyroid should not take potassium iodide unless so directed by their doctors. Also, a person should not take potassium iodide as a preventative treatment against radiation. It provides no protection in advance of radiation exposure and, in excessive amounts, can create problems of its own for the thyroid.

Words to Know

RADIOACTIVE ISOTOPE A form of an element that gives off some form of radiation and changes into another element.

FOR FURTHER INFORMATION

"Frequently Asked Questions on Potassium Iodide (KI)." U.S. Food and Drug Administration. http://www.fda.gov/cder/drugprepare/KI_Q&A.htm (accessed on November 1, 2005).

"Potassium Iodide." J. T. Baker. http://www.jtbaker.com/msds/englishhtml/p5906.htm (accessed on November 1, 2005).

"Potassium Iodide (Systemic)." Medline Plus. http://www.nlm.nih.gov/medlineplus/druginfo/uspdi/202472.html (accessed on November 1, 2005).

Wright, Jonathan V. "One Mineral Can Help a Myriad of Conditions from Atherosclerosis to 'COPD' to Zits." Tahoma Clinic. http://www.tahoma-clinic.com/iodide.shtml (accessed on November 1, 2005).

See Also Silver Iodide

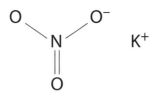

Potassium Nitrate

OVERVIEW

Potassium nitrate (poe-TAS-ee-yum NYE-trate) is transparent, colorless, or white, and may be crystalline or powdery solid. It is odorless with a sharp, cool, salty taste. It is slightly hygroscopic, that is, having a tendency to absorb moisture from the air. Potassium nitrate, more commonly known as saltpeter or niter, has been used by humans for many centuries. Going back as far as ancient Chinese civilizations, the compound was used as an ingredient in fireworks, to preserve foods, to make incense burn more evenly, to increase the male sex drive, and for magic potions.

HOW IT IS MADE

Potassium nitrate is made commercially by reacting potassium chloride (KCl) with nitric acid (HNO_3) at high temperatures: $3KCl + 4HNO_3 \rightarrow 3KNO_3 + Cl_2 + NOCl + 2H_2O$.

The compound can also be obtained for use from natural sources. It occurs as a thin, whitish, glassy crust on rocks in

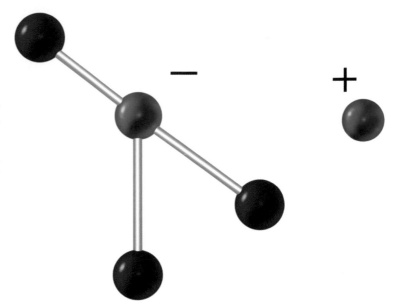

sheltered areas, such as caves. In warm climates, potassium nitrate forms when bacteria decompose animal feces and other organic matter. The compound usually appears as a white powder on the surface of soil. These sources of potassium nitrate are of use on a small scale basis and have no commercial value.

COMMON USES AND POTENTIAL HAZARDS

The primary use of potassium nitrate is in explosives, blasting powders, gunpowder, fireworks, and matches. The compound is used as an oxidizing agent in such preparations. An oxidizing agent is a substance that provides oxygen for the combustion of some other material. For example, gunpowder, the oldest known explosive, is a mixture of potassium nitrate, charcoal (nearly pure carbon), and sulfur. When the mixture is ignited, the carbon and sulfur burn very rapidly to produce carbon dioxide (CO) and sulfur dioxide (SO_2. At the same time, the potassium nitrate decomposes to produce a variety of products, one of which is nitric oxide (NO). The rapid formation of very hot gases is responsible for the shock wave produced in the explosion.

Interesting Facts

- At one time, potassium nitrate was prepared by mixing manure with mortar or wood ash, soil, and an organic material, such as straw. The bed was kept moist with urine and turned often to speed decomposition of the organic matter. After a year, the bed was thoroughly watered, dissolving the potassium nitrate that had accumulated. It was then recrystallized and purified.

- In 1862, leaders of the Confederate Army ordered a chemistry professor at South Carolina College to teach farmers how to make potassium nitrate to ensure an adequate supply of the compound for use in making gunpowder.

- Gunpowder was probably first used as early as the eleventh century. The English natural philosopher Roger Bacon (1214-1294) described a method for making gunpowder in 1242.

Some other uses of potassium nitrate include:

- As a meat preservative that helps meats retain their bright red color;

- As a flux for soldering;

- In fertilizers, especially for use with crops such as tomatoes, potatoes, tobacco, leafy vegetables, citrus fruits, and peaches;

- In the manufacture of glasses and ceramics;

- As an additive for tobacco products that helps the tobacco burn more cleanly and smoothly;

- As an oxidizing agent in rocket propulsion systems;

- As a diuretic, a substance that increases the flow of urine from the body; and

- As a raw material in the manufacture of other potassium compounds.

Exposure to moderate amounts of potassium nitrate dust and fumes can result in irritation of the skin, eyes, and respiratory system. Symptoms may include sneezing, coughing,

Words to Know

FLUX A material that lowers the melting point of another substance or mixture of substances or that is used in cleaning a metal.

dizziness, drowsiness, and headache. Ingestion of the compound may result in nausea, vomiting, and severe abdominal pain. Exposure to large quantities of the compound may have more serious consequences because it interferes with the blood's ability to transport oxygen. In such cases, a person may experience shortness of breath, bluish skin, serious damage to the kidneys, unconsciousness, and even death. People who work directly with potassium nitrate are at greatest risk for such health problems.

FOR FURTHER INFORMATION

"Material Safety Data Sheet: Potassium Nitrate." Department of Chemistry, Iowa State University. http://avogadro.chem.iastate.edu/MSDS/KNO3.htm (accessed on November 3, 2005).

Multhauf, Robert P., and Christine M. Roane. "Nitrates." In *Dictionary of American History.* Edited by Stanley I. Kutler. 3rd ed., vol. 6. New York: Charles Scribner's Sons, 2003.

"Potassium Nitrate." Hazardous Substances Data Bank. http://toxnet.nlm.nih.gov/cgi-bin/sis/search/r?dbs+hsdb:@term+@na+Potassium+Nitrate (accessed on November 3, 2005).

"Potassium Nitrate." International Chemical Safety Cards. http://www.inchem.org/documents/icsc/icsc/eics0184.htm (accessed on November 3, 2005).

"Potassium Nitrate." Skylighter. http://www.skylighter.com/potassium-nitrate.html (accessed on November 3, 2005).

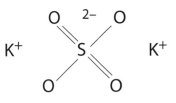

OTHER NAMES:
See Overview.

FORMULA:
K_2SO_4

ELEMENTS:
Potassium, sulfur,
oxygen

COMPOUND TYPE:
Salt (inorganic)

STATE:
Solid

MOLECULAR WEIGHT:
174.26 g/mol

MELTING POINT:
1069°C (1956°F)

BOILING POINT:
Vaporizes at 1689°C
(3072°F)

SOLUBILITY:
Soluble in water;
slightly soluble in
glycerol; insoluble in
ethyl alcohol, acet-
one, and most other
organic solvents

KEY FACTS

Potassium Sulfate

OVERVIEW

Potassium sulfate (poe-TAS-ee-yum SUL-fate) is also
known as potash of sulfur, sulfuric acid dipotassium salt,
arcanum duplicatum, and sal polychrestum. It is a colorless
or white granular, crystalline, or powdery solid with a bitter,
salty taste. It occurs in nature as the mineral arcanite and in
the mineral langbeinite ($K_2Mg_2(SO_4)_3$). The compound was
known to alchemists as early as the fourteenth century, and
was analyzed by a number of early chemists, including
Johann Glauber (1604-1670), Robert Boyle (1627-1691), and
Otto Tachenius (c. 1620-1690).

HOW IT IS MADE

A variety of methods for preparing potassium sulfate is
available. In one process, the compound is extracted from the
mineral langeinite by crushing and washing the mineral and
then separating out the double salt, $K_2Mg_2(SO_4)_3$. The product is then treated with an aqueous solution of potassium

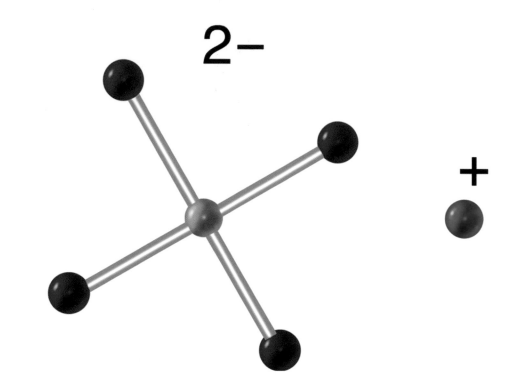

Potassium sulfate. Red atoms are oxygen; yellow atom is sulfur; and turquoise atoms are potassium. PUBLISHERS RESOURCE GROUP

chloride (KCl) to separate the two parts of the double salt from each other: $K_2Mg_2(SO_4)_3 + 4KCl \rightarrow 3K_2SO_4 + 2MgCl_2$. The compound can also be produced synthetically by treating potassium chloride with sulfuric acid (H_2SO_4): $2KCl + H_2SO_4 \rightarrow K_2SO_4 + 2HCl$.

In a variation of this procedure, potassium chloride is treated with the raw materials from which sulfuric acid is made, rather than the acid itself: $4KCl + 2SO_2 + 2H_2O + O_2 \rightarrow 2K_2SO_4 + 4HCl$.

COMMON USES AND POTENTIAL HAZARDS

Over 90 percent of the potassium sulfate produced in the United States is used as a fertilizer. It provides plants with two essential elements: potassium and sulfur. It finds its greatest use on crops that are sensitive to the chloride ion (Cl^-) present in most conventional agricultural fertilizers. Those crops include coffee, tea, tobacco, citrus fruits, grapes, and potatoes. However, its use is somewhat limited because it is twice as expensive as fertilizers that contain potassium chloride.

The second most important use of potassium sulfate is as a supplement for animal feeds, accounting for another 8 percent of the compound produced in the United States. The remaining 1 percent of potassium sulfate goes to the production of gypsum board and gypsum cement, for the synthesis of potassium alum (potassium aluminum sulfate), in the manufacturing of glass and ceramics, for the production of dyes and lubricants, and as a flash suppressant in explosives. A flash suppressent is, as its name suggests, a chemical that reduces the amount of flash produced when an explosive is detonated.

Exposure to moderate amounts of potassium sulfate appears to have no serious effects on human health. Ingestion of large amounts of the compound, can cause severe gastrointestinal irritation that requires medical attention.

Words to Know

AQUEOUS SOLUTION A solution that consists of some material dissolved in water.

SYNTHESIS A chemical reaction in which some desired chemical product is made from simple beginning chemicals, or reactants.

FOR FURTHER INFORMATION

"Potassium Sulfate." Hummel Croton, Inc.
http://www.hummelcroton.com/msds/k2so4_m.html (accessed on November 3, 2005).

"Potassium Sulfate." International Programme on Chemical Safety.
http://www.inchem.org/documents/icsc/icsc/eics1451.htm (accessed on November 3, 2005).

"Potassium Sulfate." J. T. Baker.
http://www.jtbaker.com/msds/englishhtml/p6137.htm (accessed on November 3, 2005).

See Also Potassium Carbonate, Potassium Chloride, Potassium Nitrate

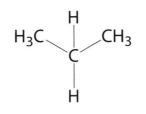

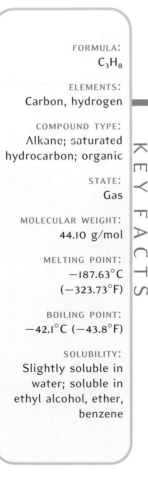
Propane

OVERVIEW

Propane (PRO-pane) is a colorless gas with an odor of natural gas. It occurs naturally in petroleum and natural gas. It belongs to the alkane family of organic compounds, compounds consisting of only carbon and hydrogen, all joined by single bonds. Propane is commonly sold as fuel, often available in a liquefied form known as liquid propane gas, or LPG.

HOW IT IS MADE

Propane is most widely available as a component of petroleum and natural gas, fossil fuels that formed many millions of years ago when marine organisms died, sank to the bottom of seas, and were eventually buried under massive layers of debris. The decay of those organisms without access to oxygen resulted in the formation of so-called fossil fuels: natural gas, petroleum, and coal. All fossil fuels are complex mixture of some free carbon and a very large variety of hydrocarbons. Natural gas, for example, consists primarily of methane, ethane, and propane.

Roughly half of all the propane produced in the United States comes from petroleum gases produced during the refining of crude oil and half from natural gas. The mixture

Interesting Facts

Propane's natural odor is so faint that it can often not be detected when leaks occur. To avoid the problem of unexpected fires and explosions, manufacturers usually add a compound with a strong odor, such as ethanediol, to make leaks more noticeable.

of gases from either refined petroleum or natural gas is lique-fied and then allowed to boil off, changing back to the original gases. Each gas boils off at its characteristic boiling point and can be captured and removed as it escapes from the liquid mixture. Often, propane is allowed to remain in its liquid state and is then made available as liquid propane gas (LPG). LPG is easier to store and transport than is gaseous propane.

COMMON USES AND POTENTIAL HAZARDS

About 100 billion liters (27 billion gallons) of propane were produced in the United States in 2004. The largest portion of that output (45 percent) was used for space heat-ing, water heating, the operation of appliances, and other purposes in commercial and residential buildings. Propane has many advantages as a fuel:

- It produces less pollution than gasoline or coal.

- Propane water heaters are less expensive to purchase and operate than are electric water heaters, and they can heat more water than electric heaters.

- Propane fireplaces are less expensive and less polluting than wood-burning fireplaces, and they can be turned on and off with a switch.

- Many professional cooks prefer propane stoves to elec-tric stoves because they heat instantly and are easier to control.

- Propane dryers take three-quarters of the time of elec-tric dryers to dry the same amount of clothes.

- Propane appliances continue to operate during power outages, unlike electric appliances.

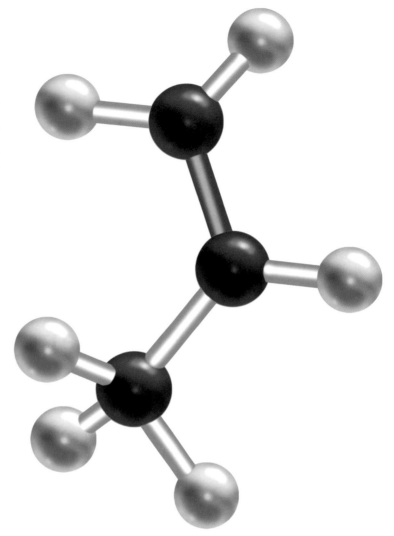

Propylene. White atoms are hydrogen and black atoms are carbon. Gray sticks indicate double bonds. PUBLISHERS RESOURCE GROUP

undergoing any change in its own chemical structure. Thermal cracking is also known as steam cracking because steam is used to produce the high temperatures need to bring about the cracking reactions. Propylene can also be prepared by the catalytic dehydrogenation of propane (C_3H_8). Catalytic dehydrogenation is a process by which hydrogen atoms are removed from a substance, resulting in the formation of double bonds where single bonds previously existed.

Interesting Facts

- Plants emit small amounts of propylene gas naturally.

- Small amounts of propylene are also produced when organic matter burns and is found in products such as cigarette smoke, automobile exhaust, and burning leaves.

COMMON USES AND POTENTIAL HAZARDS

About 15.3 million metric tons (16.8 million short tons) of propylene were produced for commercial sale in the United States in 2004. About 39 percent of that amount was used for the production of polypropylene. Almost all of the remaining production was also used for the synthesis of chemical compounds, especially acrylonitrile (14 percent), propylene oxide (11 percent), cumene (10 percent), oxo alcohols (8 percent), isopropyl alcohol (7 percent), oligomers (5 percent) and acrylic acid (3 percent). Acrylonitrile, propylene oxide, oxo alcohols, and acrylic acids are all used primarily for the production of various types of polymers. Cumene is itself used as a raw material in the production of other organic compounds, especially acetone and phenol.

In addition to the propylene produced for commercial uses, large amounts of the gas are retained by the petroleum industry for conversion to products that are added to gasoline and other petroleum-based products. According to some estimates, between half and three-quarters of all propylene that comes from cracking reactions is retained for this purpose, leaving the remainder for commercial sale.

In addition to the hazard it poses as a flammable gas, propylene does have some health risks. At relatively high concentrations, it can act as an asphyxiant, a substance that can produce unconsciousness, and can cause death. It can also act as a mild anesthetic.

Words to Know

CATALYST A material that increases the rate of a chemical reaction without undergoing any change in its own chemical structure.

POLYMER A compound consisting of very large molecules made of one or two small repeated units called monomers.

FOR FURTHER INFORMATION

"Chronic Toxicity Summary: Propylene." California Office of Environmental Health Hazard Assessment. http://www.oehha.org/air/chronic_rels/pdf/115071.pdf (accessed on November 3, 2005).

Meikle, Jeffrey L. *American Plastic: A Cultural History.* Piscataway, NJ: Rutgers University Press, 1997.

"Propylene." Equistar Chemical Company. http://www.equistarchem.com/html/petrochemical/olefins/propylene.htm (accessed on November 3, 2005).

"Propylene." LG Petrochemical. http://www.lgpetro.com/eng/product_info/product/propylene.html (accessed on November 3, 2005).

See Also Ethylene; Polypropylene

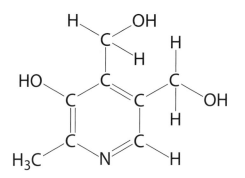

OTHER NAMES:
See Overview.

FORMULA:
CH$_3$C$_5$HN(OH)(-CH$_2$OH)$_2$

ELEMENTS:
Carbon, hydrogen, nitrogen, oxygen

COMPOUND TYPE:
Organic

STATE:
Solid

MOLECULAR WEIGHT:
169.18 g/mol

MELTING POINT:
159°C–162°C (318°F–324°F)

BOILING POINT:
Not applicable; sublimes above melting point

SOLUBILITY:
Very soluble in water; slightly soluble in ethyl alcohol and acetone

KEY FACTS

Pyridoxine

OVERVIEW

Pyridoxine (peer-ih-DOCK-seen) is also known as 3-Hydroxy-4,5-bis(hydroxymethyl)-2-methylpyridine; 3-hydroxy-4,5-dimethylol-2-methylpyridine; and vitamin B$_6$. It is a white, odorless, crystalline compound with a slightly bitter taste. The term pyridoxine is also used as a generic term for three compounds with biological activity classified under the term Vitamin B$_6$. The three compounds are pyridoxine, pyridoxal, and pyridoxamine. Pyridoxine is usually produced commercially as the hydrochloride, CH$_3$C$_5$HN(OH)(CH$_2$OH)$_2$·HCl, which has somewhat different physical characteristics from pyridoxine itself.

Vitamin B$_6$ was discovered in 1938 by five groups of researchers working independently. The five groups were all looking for a cure for a disease in rats called acrodynia that resembles the human disease pellagra. In the early 1930s, the Hungarian-American biochemist Albert Szent-Györgi (1893–1986) hypothesized the existence of a vitamin that would cure acryodynia and even gave the vitamin a

Words to Know

CATALYST A material that increases the rate of a chemical reaction without undergoing any change in its own chemical structure.

METABOLISM The process including all of the chemical reactions that occur in cells by which fats, carbohydrates, and other compounds are broken down to produce energy and the compounds needed to build new cells and tissues.

SYNTHESIS Chemical reaction in which some desired chemical product is made from simple beginning chemicals, or reactants.

subject of articles in medical journals. Other reported instances of vitamin B_6 deficiency disease have involved pregnant women who did not receive enough of the vitamin in their daily diets and people living in Cuba during the early 1990s who had restricted diets. Symptoms of vitamin B_6 deficiency include general weakness, anemia, cracked lips, inflamed tongue and mouth, irritability, depression, and skin disorders.

Meats have the highest concentration of vitamin B_6, so vegetarians may be at risk for deficiency disorders. Other foods that contain high concentrations of the vitamin include bananas, mangoes, avocados, and potatoes. Increased doses of vitamin B_6 are sometimes used to treat morning sickness and insomnia, and some authorities recommend the vitamin to decrease the risk of heart disease. The maximum recommended dose of vitamin B_6 is 50 milligrams a day.

FOR FURTHER INFORMATION

Brody, Tom. *Nutritional Biochemistry.* San Diego: Academic Press, 1998.

"Dietary Supplement Fact Sheet: Vitamin B_6." NIH Office of Dietary Supplements.
http://ods.od.nih.gov/factsheets/vitaminb6.asp (accessed on November 3, 2005).

Turner, Judith. "Pyridoxine." In *Gale Encyclopedia of Alternative Medicine.* Detroit: Gale Group, 2004.

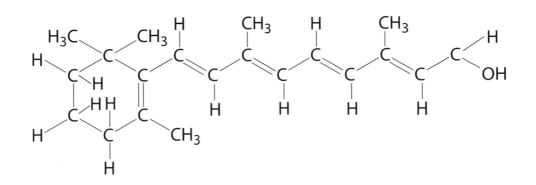

OTHER NAMES:
Vitamin A

FORMULA:
$C_{20}H_{30}O$

ELEMENTS:
Carbon, hydrogen,
oxygen

COMPOUND TYPE:
Alcohol (organic)

STATE:
Solid

MOLECULAR WEIGHT:
286.45 g/mol

MELTING POINT:
63.5°C (146°F)

BOILING POINT:
Not applicable;
decomposes

SOLUBILITY:
Practically insoluble
in water; soluble in
ethyl alcohol and
methyl alcohol

KEY FACTS

Retinol

OVERVIEW

Retinol (RET-uh-nol) is the scientific name for vitamin A, a vitamin found only in animals. It occurs as a yellowish to orange powder with a slight brownish cast and is a relatively stable compound. Retinol is converted in the body from an alcohol to the corresponding aldehyde, retinal ($C_{20}H_{28}O$), one of the primary chemical compounds involved in the process by which light is converted to nerve impulses in the retina of the eye. Vitamin A is also required for a number of other biochemical reactions in the body, including growth and development of tissue and maintenance of the immune system.

HOW IT IS MADE

Vitamin A is synthesized in animal bodies through a variety of pathways. One important source of vitamin A is a group of related compounds called the carotenes, substances responsible for the yellowish or orangish appearance of

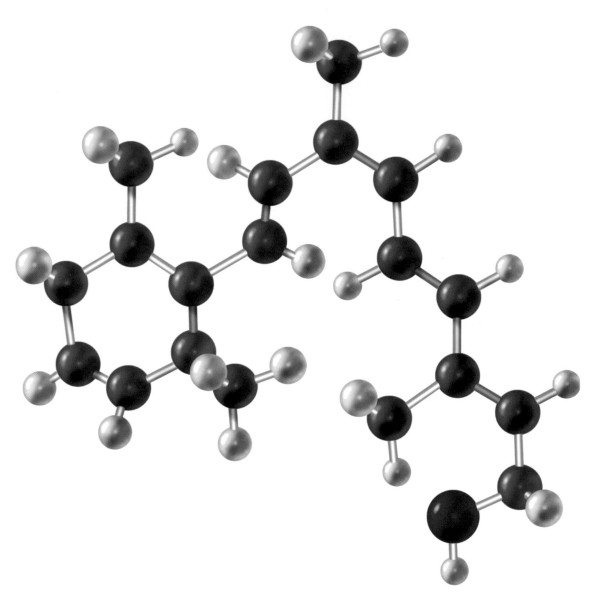

Retinol. Red atom is oxygen; white atoms are hydrogen; and black atoms are carbon. Gray sticks indicate double bonds.

fruits and vegetables such as carrots, sweet potatoes, squash, cantaloupe, apricots, pumpkin, and mangos. Some leafy green vegetables, such as collard greens, spinach, and kale, are also good sources of the carotenes. The most important of the carotenes is β-carotene (beta-carotene), $C_{40}H_{56}$. The oxidation of carotenes in animal bodies converts them to retinol.

The chemical structure of retinol was determined in 1931 by Swiss chemist Paul Karrer (1889–1971), and the compound was first prepared synthetically shortly thereafter by Austrian-German chemist Richard Kuhn (1900-1967). The first successful process for producing retinol commercially was developed in the mid-1940s by German chemist Otto Isler (1920-1992), then employed at the pharmaceutical company Roche, located in Sissein, Germany. Isler's process involved a complex series of reactions that begins with the combination of a fourteen carbon hydrocarbon and a six carbon hydrocarbon to create the fundamental backbone from which the retinol molecule is constructed. Regular production of vitamin A began in 1948 with a projected output of 10 kilograms per month, which before long was raised to 50 kilograms per month. The Roche plant at Sissein continues to produce retinol today.

COMMON USES AND POTENTIAL HAZARDS

Vitamin A is probably best known for its role in maintaining normal vision. Deficiencies of the compound are likely to manifest themselves earliest in a variety of eye problems, most commonly night blindness. Night blindness is a condition in which one loses the ability to distinguish objects in reduced light. If left untreated, vitamin A deficiencies may lead to decreased ability to see in normal light and, eventually, to complete blindness.

But vitamin A has been shown to have a number of other functions in the body. It is essential for the maintenance of growth, bone formation, reproduction, proper immune system function, and healing of wounds. A number of additional claims have been made for the compound, although evidence is not as strong as it is for the above functions. For example,

Words to Know

CARBOHYDRATE A group of sugars and starches produced by plants and used as food.

JAUNDICE A disease caused by malfunction of the liver in which the skin and eyes become yellow.

LEGUME A member of the pea family of plants, which includes peas and beans.

METABOLISM The process that includes all of the chemical reactions that occur in cells by which fats, carbohydrates, and other compounds are broken down to produce energy and the compounds needed to build new cells and tissues.

MUCOUS MEMBRANES Tissues that line the moist inner lining of the digestive, respiratory, urinary and reproductive systems.

SOLVENT A substance that is able to dissolve one or more other substances.

SYNTHESIS A chemical reaction in which some desired chemical product is made from simple beginning chemicals, or reactants.

In combination with vitamin A, riboflavin helps maintain the mucous membranes that line the digestive tract. Pregnant women need riboflavin to help the fetus grow and develop. The vitamin is also essential for eye health. Some medical professionals recommend riboflavin for the treatment of eye disorders, such as cataracts, sensitivity to bright light, and bloodshot, burning, or itching eyes. Doctors have begun experimenting with the use of riboflavin to treat migraine headaches.

In spite of its many important functions in the body, riboflavin deficiencies do not lead to serious medical problems or death. They may result in a delayed healing of wounds or relatively minor medical conditions such as cheilosis (a reddening and soreness in the corners of the lips), angular stomatitis (cracking of the skin at the corner of the lips), dermatitis (an oily skin disorder), glossitis (a swollen, reddened tongue), and cracking around the nose. In all such cases, however, a riboflavin deficiency by itself is almost never the cause of the problem; instead, deficiencies of other vitamins in the B group are also involved.

Riboflavin deficiencies are very rarely seen in the United States because the vast majority of people consume a diet that contains adequate amounts of the vitamin. Individuals most likely to suffer from riboflavin deficiency problems are those with anorexia (a condition in which people refuse to eat adequate amounts of food), older people with poor diets, alcoholics (because alcohol impairs a person's ability to absorb and use the vitamin), and newborn babies being treated for jaundice by exposure to ultraviolet light (because light destroys riboflavin).

FOR FURTHER INFORMATION

"Food Safety: From the Farm to the Fork." European Commission Web Site.
http://europa.eu.int/comm/food/fs/sc/scf/out18_en.html (accessed on November 3, 2005).

"Riboflavin." Linus Pauling Institute, Micronutrient Information.
http://lpi.oregonstate.edu/infocenter/vitamins/riboflavin/ (accessed on November 3, 2005).

"Riboflavin." MedlinePlus.
http://www.nlm.nih.gov/medlineplus/ency/article/002411.htm (accessed on November 3, 2005).

"Vitamin B2 (Riboflavin)." Herb & Supplement Encyclopedia.
http://www.florahealth.com/flora/home/canada/healthinformation/encyclopedias/VitaminB2.asp (accessed on November 3, 2005).

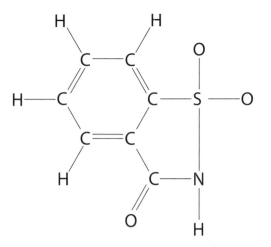

OTHER NAMES:
Benzoylsulfonic
imide; benzoic sulfi-
mide

FORMULA:
$C_7H_5NO_3S$

ELEMENTS:
Carbon, hydrogen,
nitrogen, oxygen, sul-
fur

COMPOUND TYPE:
Organic

STATE:
Solid

MOLECULAR WEIGHT:
183.18 g/mol

MELTING POINT:
228°C (442°F)

BOILING POINT:
Not applicable;
decomposes

SOLUBILITY:
Slightly soluble in
water; soluble in
acetone and ethyl
alcohol

KEY FACTS

Saccharin

OVERVIEW

Saccharin (SAK-uh-rin) is a synthetic compound whose water solutions are at least 500 times as sweet as table sugar. It passes through the human digestive system without being absorbed, so it has an effective caloric value of zero. It is used as a sugar substitute by diabetics or by anyone wishing to reduce their caloric intake.

Saccharin was the first artificial sweetener discovered. It was synthesized accidentally in 1879 when Johns Hopkins researchers Constantine Fahlberg (1850-1910) and Ira Remsen (1846-1927) were working on the development of new food preservatives. The story is told that Fahlberg accidentally spilled one of the substances being studied on his hand. Some time later, he noticed the sweet taste of the substance and began to consider marketing the product as an artificial sweetener. Fahlberg and Remsen jointly published a paper describing their work, but Fahlberg, without Remsen's knowledge, went on to request a patent for the discovery. He eventually became very wealthy from proceeds of the discovery, none of

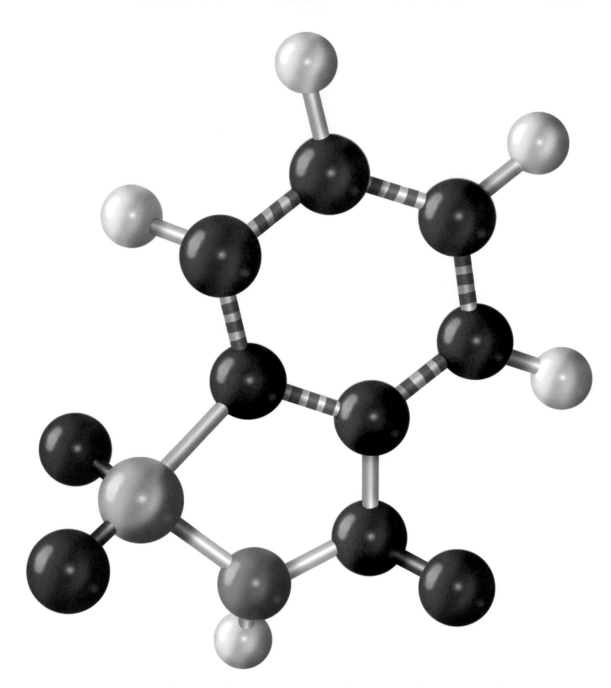

Benzoic sulfinide. Red atoms are oxygen; white atoms are hydrogen; black atoms are carbon; blue atom is nitrogen; and yellow atom is sulfur. Gray sticks indicate double bonds. Striped sticks show a benzene ring.
PUBLISHERS RESOURCE GROUP

which he shared with Remsen. Remsen was later quoted as saying that "Fahlberg is a scoundrel. It nauseates me to hear my name mentioned in the same breath with him."

In spite of its sweet taste, saccharin was used at first as an antiseptic, a substance that stops the growth of or kills bacteria. Before long, however, its primary use became that of an artificial sweetener. By 1902, it had become so popular in Germany that the German sugar industry lobbied for laws limiting production of saccharin. Similar actions occurred in the United States in 1907 and, by 1911, the federal government restricted use of the compound to overweight invalids.

A shortage of sugar during World War I (1914-1918) led to the reintroduction of saccharin as a sweetening agent in foods. Another sugar shortage during World War II (1939-1945) saw a new boom in saccharin production. This time, the compound's popularity continued after the war ended.

Questions about saccharin's safety have been raised a number of times in the past. In 1969, for example, the sweetener cyclamate was found to be carcinogenic, and its use was banned in the United States. Doubts over saccharin's safety resurfaced, partly since it was often mixed with cyclamate in artificial sweeteners. In 1972, studies with rats suggested that saccharin too might be carcinogenic, and the U.S. Food and Drug Administration imposed restrictions on the sweetener's use. Later studies attempting to reproduce the 1972 research were largely unsuccessful, and the status of saccharin as a carcinogen remain unsettled.

In 1977, the Canadian government decided that sufficient evidence existed to ban the use of saccharin except for use with diabetics and others with special medical problems. The U.S. government considered taking similar action, but, after more than a decade of reviewing the evidence, decided to allow the use of saccharin among the general public. Nonetheless, the status of saccharin as a potential health hazard remains the subject of an active debate in the United States and other parts of the world.

HOW IT IS MADE

A number of methods are available for the synthesis of saccharin. For many years, the most popular process was one developed by the Maumee Chemical Company of Toledo, Ohio, in 1950. This method begins with anthranilic acid (o-aminobenzoic acid; $C_6H_4(NH_2)COOH$), which is treated successively with nitrous acid (HNO_2), sulfur dioxide (SO_2), chlorine (Cl_2), and ammonia (NH_3) to obtain saccharin. Another process discovered in 1968 starts with o-toluene, which is then treated with sulfur dioxide and ammonia to obtain saccharin.

Interesting Facts

- The chemical company Monsanto was founded in 1901 for the sole purpose of making saccharin. The company sold all of its product to a single company, the Coca-Cola company, founded in the late 1880s.

Saccharin is not very soluble, so it is commonly made into its sodium or calcium salt (sodium saccharin or calcium saccharin), both of which readily dissolve in water, for use in drinks and cooking. Saccharin is often blended with other sweeteners to reduce its metallic aftertaste.

COMMON USES AND POTENTIAL HAZARDS

Saccharin is used almost exclusively as an artificial sweetener in food and drinks to replace sugar. Its lack of calories makes it suitable for diet products and for medical preparations designed for people who must reduce their caloric intake. It also finds some small application as a food preservative, as an antiseptic agent, and as a brightening agent in electroplating procedures.

Raw saccharin can be an irritant to the skin, eyes, and respiratory system. If ignited, it burns with the release of irritating fumes. Only individuals who come into contact

Words to Know

CARCINOGEN A substance that causes cancer in humans or other animals.

ELECTROPLATING A process by which a thin layer of one metal is deposited on top of a second metal by passing an electric current through a solution of the first metal.

SALT A compound in which all of the hydrogen ions are replaced by metal ions.

SYNTHESIS A chemical reaction in which some desired chemical product is made from simple beginning chemicals, or reactants.

with large quantities of saccharin are likely to be concerned about such safety problems, however.

FOR FURTHER INFORMATION

Henkel, John. "Sugar Substitutes: Americans Opt for Sweetness and Lite." *FDA Consumer.* December 1999. Available online at http://www.fda.gov/fdac/features/1999/699_sugar.html (accessed on November 3, 2005).

Nabors, Lyn O'Brien, ed. *Alternative Sweeteners (Food Science and Technology).* Third revision. London: Marcel Dekker, 2001.

Ruprecht, Wilhelm. "Consumption of Sweeteners: An Evolutionary Analysis of Historical Development." http://www.druid.dk/conferences/nw/paper1/ruprecht.pdf (accessed on November 3, 2005).

"Saccharin Sodium." J. T. Baker. http://www.jtbaker.com/msds/englishhtml/s0073.htm (accessed on November 3, 2005).

$$\begin{array}{c} O \\ \| \\ Si \\ \| \\ O \end{array}$$

OTHER NAMES:
Silica, quartz, sand, amorphous silica, silica gel, and others

FORMULA:
SiO_2

ELEMENTS:
Silicon, oxygen

COMPOUND TYPE:
Nonmetallic oxide (inorganic)

STATE:
Solid

MOLECULAR WEIGHT:
60.08 g/mol

MELTING POINT:
Varies depending on crystalline state; typically above 1700°C (3100°F)

BOILING POINT:
2950°C (5300°F)

SOLUBILITY:
Solubility depends on crystalline state; generally insoluble in water; soluble in many acids and alkalis

KEY FACTS

Silicon Dioxide

OVERVIEW

Silicon dioxide (SILL-uh-kon dye-OK-side) is one of the most abundant chemical compounds on Earth. It makes up about 60 percent of the weight of the Earth's crust either as an independent compound (SiO_2) or in combination with metallic oxides that form silicates. Silicates are inorganic compounds whose negative part is the SiO_3^- ion (grouping of atoms). An example is magnesium silicate, $MgSiO_3$.

Silicon dioxide occurs as colorless, odorless, tasteless white or colorless crystals or powder. Its many different forms can be classified as crystalline, amorphous, or vitreous. In crystalline forms of silicon dioxide, all of the atoms that make up the substances are arranged in orderly patterns that have the shape of cubes, rhombohedrons, or other geometric figures. In amorphous silicon dioxide, silicon and oxygen atoms are arranged randomly, without any clear-cut pattern. Vitreous silicon dioxide is a glassy form of the compound that may be transparent, translucent, or opaque. The various forms of silicon dioxide can be converted from one form to another by heating and changes in pressure.

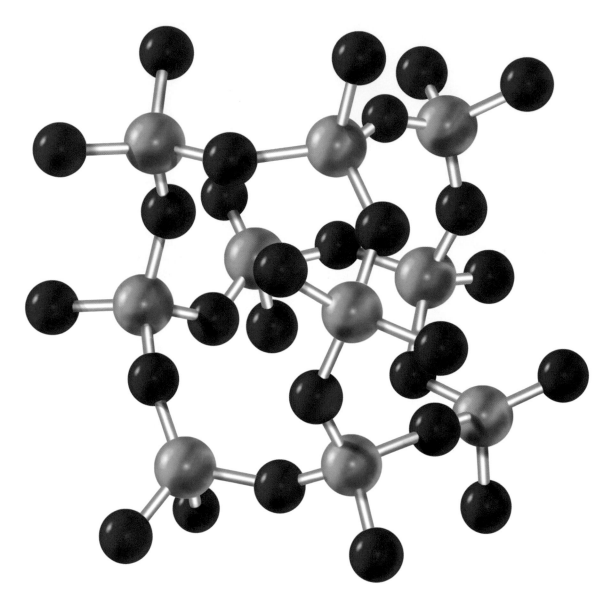

Silicon dioxide. Red atoms are oxygen and yellow atoms are silicon. PUBLISHERS RESOURCE GROUP

An especially interesting form of silicon dioxide is silica gel, a powdery form of amorphous silicon dioxide that is highly adsorbent. An adsorbent material (in contrast to an absorbent material) is one that is capable of removing a material, such as water, ammonia, alcohol, or other gases, out of the air. The second material bonds weakly to the outer surface of silica gel particles. Silica gel is able to adsorb anywhere from 30 to 50 percent of its own weight in water

CHEMICAL COMPOUNDS

Interesting Facts

- Stardust, a U.S. National Aeronautics and Space Administration (NASA) spacecraft, used silica gel to collect particles of debris from the tail of comet Wild-2.

- Although silica gel has been known since the mid-seventeenth century, practical applications for the material were not discovered until 1919 when American Walter A. Patrick (1888–1969) patented a number of processes for the manufacture of the compound. It still did not become widely popular until World War II (1939–1945), when the American military found a number of important uses for the compound.

from the surrounding atmosphere before it becomes saturated. The silica gel is not chemically altered by the process of adsorption and still feels dry even when saturated. The adsorbed water can be driven off simply by heating the silica gel, allowing the material to regain its adsorbent properties.

HOW IT IS MADE

Although methods are available for synthesizing silicon dioxide, there is no practical reason for doing so. The abundant quantities of silicon dioxide found in the earth's crust are sufficient to satisfy all industrial needs. Among the minerals and earths that contain silicon dioxide in an uncombined form are quartz, flint, diatomite, stishovite, agate, amethyst, chalcedony, cristobalite, and tridymite.

Naturally occurring silicon dioxide can be treated by a variety of physical processes to change its form. For example, heating crystalline silicon dioxide above its melting point and then cooling it again converts the compound into its vitreous form, sometimes called natural glass. Silica gel is made by treating sodium silicate (Na_2SiO_3) with sulfuric acid (H_2SO_4): $Na_2SiO_3 + H_2SO_4 \rightarrow SiO_2 + H_2O + Na_2SO_4$.

COMMON USES AND POTENTIAL HAZARDS

The primary use of silicon dioxide is in the building industry. It is used to make ceramics, enamels, concrete, and specialized silica bricks used as refractory materials. It is also one of the raw materials from which all kinds of glass are made. Vitreous silicon dioxide is an important constituent of specialized types of glass, such as that used in making laboratory equipment, mirrors, windows, prisms, cells, and other kinds of optical devices. Silicon dioxide is also used as an anti-caking or thickening agent in a variety of foods and pharmaceutical products. Some other applications of silicon dioxide include:

- In the manufacture of polishing and grinding materials;
- As molds for casting;
- In the production of elemental silicon;
- As a filler in many different kinds of products, including paper, insecticides, rubber products, pharmaceuticals, and cosmetics;
- As an additive in paints to produce a low-gloss finish;
- In the reinforcement of certain types of plastics.

The primary application of silica gel is as a drying agent. Packets of silica gel are found in many consumer products, such as electronic equipment, hardware tools, clothing, CD and DVD discs, and foodstuffs. Because of its ability to adsorb moisture from the surrounding air, silica gel prevents rust and other forms of oxidation. Silica gel also has similar applications in industry. For example, it is used to dry compressed air, air conditioning systems, and natural gas. The compound is also used to bleach petroleum oils and as an anti-caking agent for cosmetics and pharmaceuticals.

Normal exposure to silicon dioxide is not considered hazardous to the health of humans or other animals. Inhaling silicon dioxide dust or fumes, however, may create problems in the respiratory system. In low doses, inhaled silica can cause coughing, wheezing, and difficulty breathing. In higher doses, particles of silicon dioxide may lodge in the lungs and block the openings through which oxygen is absorbed. Over long periods of exposure, a condition known as silicosis may develop. Silicosis is a condition similar to tuberculosis, lung cancer, or emphysema in which a person's ability to breathe normally is reduced, causing severe and

Words to Know

ALKALI A strong base.

REFRACTORY A material with a high melting point, resistant to melting,

often used to line the interior of industrial furnaces.

life-threatening long-term health problems. Individuals at greatest risk for silicosis and other silicon dioxide-related problems are those who cut, chip, drill, or grind objects that contain silica. During these processes, silica is reduced to a fine powder that is easily inhaled. Wearing a mask during these operations is generally sufficient to protect a worker from inhaling these fumes and particles.

FOR FURTHER INFORMATION

Brown, David. "Walter Patrick: Preservationist of the First Order." *Mount Washington Newsletter.* Spring 2003. Also available online at http://www.mwia.org/Newsletters/MWIANewsletter_Spring%202003.pdf (accessed on December 10, 2005).

Brownlee, Don. "An Exciting Encounter with a Cold Dark Mysterious Body from the Edge of the Solar System." Jet Propulsion Laboratory. California Institute of Technology. http://stardust.jpl.nasa.gov/news/news101.html (accessed on December 10, 2005).

"Crystalline Silica Exposure." U.S. Occupational Safety and Health Administration. http://www.osha.gov/Publications/osha3177.pdf (accessed on December 10, 2005).

Scheer, James F. "Silica: Health and Beauty from Nature." *Better Nutrition* (December 1997): 38-43. Also available online at http://www.findarticles.com/p/articles/mi_m0FKA/is_n12_v59/ai_20086185 (accessed on December 10, 2005).

"Silica Gel." J. T. Baker. http://www.jtbaker.com/msds/englishhtml/s1610.htm (accessed on December 10, 2005).

"Silica—Silicon Dioxide." Azom.com. http://www.azom.com/details.asp?ArticleID=1114 (accessed on December 10, 2005).

"What Is Silica Gel and Why Do I Find Packets of It in Everything I Buy?" How Stuff Works. http://science.howstuffworks.com/question206.htm (accessed on December 10, 2005).

See Also Calcium Silicate; Sodium Silicate

KEY FACTS

OTHER NAMES:
Silver(I) iodide

FORMULA:
AgI

ELEMENTS:
Silver, iodine

COMPOUND TYPE:
Binary salt (inorganic)

STATE:
Solid

MOLECULAR WEIGHT:
234.77 g/mol

MELTING POINT:
558°C (1036°F)

BOILING POINT:
1506°C (2743°F)

SOLUBILITY:
Insoluble in water and organic solvents; soluble in solutions of sodium chloride, potassium chloride, and ammonium hydroxide

Silver Iodide

OVERVIEW

Silver iodide (SILL-ver EYE-oh-dide) is a light yellow crystalline or powdery material that darkens on exposure to light. The darkening occurs because silver ions (Ag^+; silver atoms with a positive charge) are converted to neutral silver atoms (Ag^0) that are dark gray in color. Silver iodide is used primarily in photography and in cloud-seeding experiments.

HOW IT IS MADE

Silver iodide occurs naturally in the mineral odargyrite (also known as iodyrite), from which it can be extracted. The compound is extracted by adding concentrated hydriodic acid (HI) to the mineral, which dissolves out the silver iodide. The silver iodide can then be recovered by evaporating the solution and purifying the product. The compound can also be made more easily by reacting silver nitrate ($AgNO_3$) with sodium or potassium iodide (NaI or KI) under a ruby red light. The colored light is used to prevent oxidation of the

Silver iodide. Turqouise atom is sliver and yellow atom is iodine. PUBLISHERS RESOURCE GROUP

silver ion to silver metal during the preparation: $AgNO_3 + NaI \rightarrow AgI + NaNO_3$.

COMMON USES AND POTENTIAL HAZARDS

The primary use of silver iodide is in photography. The compound was first used for this purpose in the early nineteenth century by the French experimenter Louis Jacques Mandé Daguerre (1787-1851). Daguerre first covered a sheet of copper metal with a thin layer of silver. He then exposed the plate to iodine vapors, converting the silver metal to silver iodide. The photograph was taken by exposing the plate to light, which converted colorless silver ions in silver iodide back to grayish silver atoms. After exposure, the plate was treated with magnesium vapor, which adhered to the parts of the plate that had been exposed to light. The unreacted silver iodide was then washed off the plate, revealing an image on the plate. The fundamental principle invented by Daguerre is still used today in taking photographs with photograph film containing silver iodide.

The other major use for silver iodide is in cloud seeding. Cloud seeding is the process by which some foreign material—usually silver iodide or dry ice (solid carbon dioxide)—is dropped into a rain cloud. The crystals of silver iodide or carbon dioxide provide nuclei—tiny cores—on which water can condense to form water droplets. The process of cloud seeding was first developed in the 1940s by American chemist Vincent Schaefer (1927-1993). Schaefer used crushed

Interesting Facts

- The United States government spent about $19 million per year in the 1970s for cloud seeding experiments. By the late 1990s, that amount had been reduced to about $500,000 as doubts were raised about the effectiveness of the procedure.

dry ice dropped from an airplane into a cloud in his first experiments. His initial experiments proved to be successful, with rain or snowstorms resulting from this seeding.

For three decades, researchers attempted to modify and improve Schaefer's techniques so as to be able to use cloud seeding to produce rain on demand. One step in that direction was the effort by General Electric scientist Bernard Vonnegut (1914-1997) to use silver iodide crystals rather than dry ice for seeding. Although these experiments produced moderately successful results, many scientists eventually lost confidence in the process of cloud seeding as a reliable method for producing rain. By the early twenty-first century, the process was being used on only a small scale in isolated areas in efforts to produce rain. The problem is that assessing the success of cloud seeding is very difficult. Because the process requires that clouds are already in an area, it is difficult to determine whether rain or snow falls as a result of cloud seeding or as a result of natural processes. In 2003, the National Academy of Sciences reached the conclusion that no reliable scientific evidence exists to suggest that cloud seeding produces more rain or snow than would occur naturally. The American Meteorological Society has taken a somewhat more optimistic view, suggesting that cloud seeding may increase precipitation by up to 10 percent.

Some concerns have been expressed about the environmental and health effects of using silver iodide for cloud seeding. However, only small amounts of silver iodide are released into the atmosphere. That which does fall to earth does not dissolve in water and so is unlikely to enter a community water supply. Tests have shown that the concentration of silver iodide in rainwater is far below the 50 micrograms per liter that has been deemed safe by the U.S. Public Health Service. The primary health concern for workers who handle silver iodide is a condition known as argyreia, in which the skin is stained a bluish black color by the compound.

FOR FURTHER INFORMATION

"Chemistry of Photography." Cheresources. http://www.cheresources.com/photochem.shtml (accessed on November 3, 2005).

Fukuta, Norihiko. "Cloud Seeding Clears the Air." *Physics in Action* (May 1998). Also available online at http://physicsweb.org/articles/world/11/5/3/1 (accessed on November 3, 2005).

"Silver Iodide." ESPI Metals. http://www.espimetals.com/msds's/silveriodide.pdf (accessed on November 3, 2005).

"Silver Iodide." H&S Chemical. http://www.hschem.com/msds/silverIodide.htm (accessed on November 3, 2005).

Zertuche, Casey. "Every Cloud Has a Silver Lining." The Daily Texan (April 12, 2004). Also available online at http://www.dailytexanonline.com/media/paper410/news/2004/04/12/Focus/Every.Cloud.Has.A.Silver.Lining-657354.shtml?norewrite&sourcedomain=www.dailytexanonline.com (accessed on November 3, 2005).

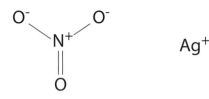

OTHER NAMES:
Silver(I) nitrate; lunar caustic

FORMULA:
$AgNO_3$

ELEMENTS:
Silver, nitrogen, oxygen

COMPOUND TYPE:
Salt (inorganic)

STATE:
Solid

MOLECULAR WEIGHT:
169.87 g/mol

MELTING POINT:
212°C (414°F)

BOILING POINT:
440°C (824°F); decomposes

SOLUBILITY:
Soluble in water, glycerol, and hot ethyl alcohol; moderately soluble in acetone

Silver Nitrate

OVERVIEW

Silver nitrate (SILL-ver NYE-trate) is a colorless to transparent to white crystalline solid with no odor and a bitter metallic taste. In pure form, the compound is not affected by light, but trace amounts of organic impurities may catalyze the conversion of silver ions (Ag^+; silver atoms with a positive charge) to grayish neutral silver atoms (Ag^0) that give the salt a grayish tint. Silver nitrate is the most widely used of all silver compounds, finding application in the synthesis of other silver compounds, as a catalyst in certain industrial chemical reactions, as an antiseptic and germicide, and in photographic processes.

The therapeutic effects of silver compounds, including silver nitrate, have been known for many centuries. Both the Greeks and Romans, for example, used aqueous solutions of silver nitrate to treat wounds and cuts. In 1881, the German physician Carl Crede (1819-1892) developed the practice of applying a 2 percent solution of silver nitrate to the eyes of newborn babies to prevent gonorrheal

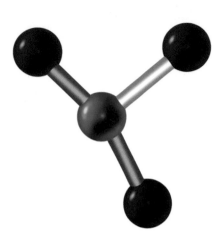

Silver nitrate. Red atoms are oxygen; blue atom is nitrogen; and turquoise atom is silver. Gray sticks indicate double bonds. PUBLISHERS RESOURCE GROUP

ophthalmia, a bacterial infection of the eyes that may result in blindness in a child.

The use of silver nitrate in printing and photography dates to discoveries made in the 1720s by the German chemist Johann Schulze (1687-1744). Schulze found that a mixture of silver, nitric acid, and chalk turns purple or black when exposed to light. That simple discovery formed the basis of the modern science of photography. In 1802, Thomas Wedgwood (1771-1805), son of the founder of the famous Wedgwood pottery company, used silver nitrate to make temporary negative prints on paper, producing shades of gray as well as pure black and white. In 1835, the British mathematician William Henry Fox Talbot (1800-1877) made the first permanent paper negative from paper coated with silver nitrate and common table salt (sodium chloride).

Two years later, the French inventor Louis Jacques Mandé Daguerre (1787-1851) coated a copper plate with silver and washed it with nitric acid to create a plate on which he made the first Daguerreotype, an early form of photography. By the 1850s, Daguerre's technique was in wide use, with the substitution of glass plates coated with silver nitrate to obtain images produced by exposing the plates to light. By the end of the nineteenth century, emulsions of silver nitrate in celluloid were being used for making photographs, a process that was modified in the 1930s by the substitution of

Interesting Facts

- Film speed numbers, such as ISO 100 or ISO 4000, correspond to the size of silver nitrate particles on the film. The finer the grain, the more detailed the image quality.

- Silver nitrate is not toxic to humans or other mammals. But it is toxic to fish and other aquatic organisms. For this reason, it should not be discarded in lakes and rivers.

cellulose acetate for celluloid. Silver nitrate remains an important component of the photographic process today.

HOW IT IS MADE

Silver nitrate is made by dissolving metallic silver in weak nitric acid.

$$2Ag + 2HNO_3 \rightarrow 2AgNO_3 + H_2$$

The solution is then evaporated to recover the crystalline silver nitrate, which is heated to remove impurities, dissolved in water, and re-purified.

COMMON USES AND POTENTIAL HAZARDS

The primary use of silver nitrate is in the production of other silvers salts used in the production of photographic film. Compounds such as silver bromide (AgBr) and silver iodide (AgI) decompose when exposed to light, forming free silver:

$$AgBr \rightarrow Ag^O + Br^- \text{ and } AgI \rightarrow Ag^O + I^-$$

While silver bromide and silver iodide are nearly colorless, the free silver atoms formed in this reaction are black. Thus, portions of a photographic film that have been exposed to light turn black where free silver has formed.

In addition to its use in treating the eyes of newborn babies, silver nitrate has a number of other medical applications. It is used as a germicidal wall spray in medical facilities, as a topical (on the skin) anti-infective agent, for the cauterization of wounds, and as a general antiseptic. Cauterization is the process by which the skin surrounding a wound

Words to Know

ANTISEPTIC A chemical that prevents the growth of bacteria and viruses that cause disease.

AQUEOUS Referring to solution that consists of some material dissolved in water.

CATALYST A material that increases the rate of a chemical reaction without

undergoing any change in its own chemical structure.

EMULSION A temporary mixture of two liquids that normally do not dissolve in each other.

THERAPEUTIC DRUG A compound that has healing properties.

is burned in order to seal the wound. Other uses of silver nitrate include:

- The silvering of mirrors, a process by which a thin coating of silver metal is attached to the back of a piece of glass to form a mirror;

- As a catalyst in the manufacture of ethylene oxide, an important raw material in the production of plastics;

- In the silver plating of metals and plastics;

- In the manufacture of indelible printing inks;

- For hair dye; and

- As a flower preservative, a process that involves adding a small amount of silver nitrate solution to the water in which flowers are stored.

Silver nitrate is an irritant that can cause inflammation and burning of the skin, eyes, and respiratory system. It is also toxic by ingestion. In sufficient amounts, silver nitrate can cause severe damage to the respiratory tract and lungs, blindness, and even death. Solutions of silver nitrate can stain the skin dark purple or black, although such stains do not necessarily indicate that serious damage has occurred.

FOR FURTHER INFORMATION

Dunn, Peter M. "Dr Carl Credé (1819-1892) and the Prevention of Ophthalmia Neonatorum." *Archives of Diseases in Childhood* (September 2000): F158-F159. Also available online at http://fn.bmjjournals.com/cgi/content/full/83/2/F158 (accessed on November 5, 2005).

Grimm, Tom, and Michele Grimm. *The Basic Book of Photography.* New York: Plume Books, 2003.

Patnaik, Pradyot. "Silver Nitrate." *Handbook of Inorganic Chemicals.* New York: McGraw-Hill, 2003, 841-842.

"Silver Nitrate." New Jersey Department of Health and Senior Services.http://nj.gov/health/eoh/rtkweb/1672.pdf (accessed on November 5, 2005).

See Also Cellulose Nitrate; Silver Iodide

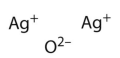

Ag^+ $\quad$ Ag^+

O^{2-}

Silver(I) Oxide

OTHER NAMES:
Silver oxide; argentous oxide

FORMULA:
Ag_2O

ELEMENTS:
Silver, oxygen

COMPOUND TYPE:
Metallic oxide (inorganic)

STATE:
Solid

MOLECULAR WEIGHT:
231.74 g/mol

MELTING POINT:
Decomposes at about 200°C (400°F)

BOILING POINT:
Not applicable

SOLUBILITY:
Slightly soluble in water; insoluble in ethyl alcohol

K E Y   F A C T S

OVERVIEW

Silver(I) oxide (SILL-ver one OK-side) is an odorless dark brown or black powder with a metallic taste. It is used primarily for polishing glass, the purification of water, and coloring glass.

HOW IT IS MADE

Silver(I) oxide is made by reacting silver nitrate ($AgNO_3$) with sodium or potassium hydroxide (NaOH or KOH). For example:

$$2AgNO_3 + 2NaOH \rightarrow Ag_2O + 2NaNO_3 + H_2O$$

The silver(I) oxide settles out as a precipitate that can then be washed and purified.

COMMON USES AND POTENTIAL HAZARDS

Silver(I) oxide finds limited commercial and industrial application. It is used as an ingredient in the manufacture of

Silver oxide. Red atom is oxygen and silver atoms are silver. PUBLISHERS RESOURCE GROUP

glass to give a yellowish caste to the glass. It is also a component of mixtures used to polish glass, including the glass used in optical lenses. Silver(I) oxide is also used as a catalyst in certain industrial operations and in some water purification systems.

Silver(I) oxide is a skin, eye, and respiratory irritant that may cause coughing, wheezing, shortness of breath, and pulmonary edema (accumulation of fluid in the lungs). It can also cause burning of the eyes and skin. Ingestion can produce burning of the gastrointestinal tract accompanied by nausea, vomiting, and abdominal pain. Long-term exposure to silver(I) oxide can cause argyreia, a bluish-gray discoloration of the skin, eyes, and mucous membranes (the soft tissues lining the breathing and digestive passages).

Words to Know

AQUEOUS Referring to a solution that consists of some material dissolved in water.

PRECIPITATE A solid material that settles out of a solution, often as the result of a chemical reaction.

FOR FURTHER INFORMATION

"Material Safety Data Sheet." IC Controls.
    http://www.iccontrols.com/files/a1100122.pdf   (accessed   on
    November 5, 2005).

"Silver(I) Oxide." Patnaik, Pradyot. *Handbook of Inorganic Chemicals.* New York: McGraw-Hill, 2003, 842-843.

"Silver Oxide." Chem007.
    http://www.chem007.com/specification_d/chemicals/supplier/
    cas/Silver%20oxide.asp (accessed on November 5, 2005).

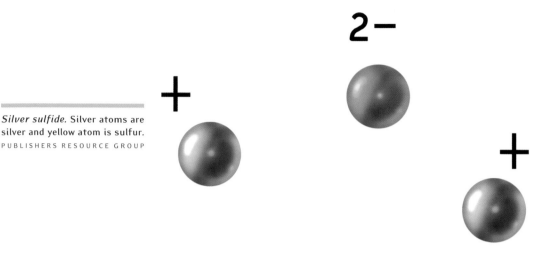

Silver sulfide. Silver atoms are silver and yellow atom is sulfur.
PUBLISHERS RESOURCE GROUP

COMMON USES AND POTENTIAL HAZARDS

Naturally occurring silver(I) sulfide can be used as a source of silver metal. The sulfide is roasted in air, converting silver(I) sulfide to silver(I) sulfate. The sulfate can then be treated chemically to obtain silver metal. The process finds little commercial application since other, more economically efficient, sources of silver are available.

The primary use of silver(I) sulfide is in the production of glazes for ceramics. The compound gives a glassy or metallic sheen to the glaze. It is also used in the process known as niello, first used by the early Greeks and Romans to produce a black, metallic inlay on the surface of pottery.

Silver(I) sulfide is an irritant to the skin, eyes, and respiratory system. No long-term effects of exposure to the compound have been reported. When absorbed through cuts in the skin or ingested it may produce a condition known as argyreia, which causes skin and mucous membranes to develop a bluish-black color.

FOR FURTHER INFORMATION

Krampf, Robert. "Cleaning the Silver." Edgerton Explorit Center. http://www.edgerton.org/kidscorner/silver.html (accessed on November 5, 2005).

"Material Safety Data Sheet." ESPI Metals. http://www.espimetals.com/msds's/silversulfide.pdf (accessed on November 5, 2005).

Interesting Facts

- Tarnish is caused by a chemical reaction between silver in tableware and sulfur compounds that occur in eggs. Tarnish can be removed by soaking tableware in a solution of warm baking soda in water in a pan lined with aluminum foil. A chemical reaction occurs in which aluminum replaces silver, forming aluminum sulfide (Al_2S_3) and free silver: $2Al + 3Ag_2S \rightarrow 6Ag + Al_2S_3$. The reaction occurs only if the silver tableware is in physical contact with the aluminum foil because an electric current must flow between the two metals.

- The silver present in electric contacts exposed to hydrogen sulfide gas may begin to grow long filaments known as *silver whiskers*. These whiskers have been known to grow as long as 8 centimeters (3 inches) in length and can cause catastrophic failures in the electrical contacts.

"Mineral Acanthite/Argentite, The." Amethyst Galleries. http://mineral.galleries.com/minerals/sulfides/acanthit/acanthit.htm (accessed on November 5, 2005).

Patnaik, Pradyot. *Handbook of Inorganic Chemicals.* New York: McGraw-Hill, 2003, 845.

Words to Know

AQUEOUS Referring to a solution that consists of some material dissolved in water.

PRECIPITATE A solid material that settles out of a solution, often as the result of a chemical reaction.

SYNTHESIS A chemical reaction in which some desired chemical product is made from simple beginning chemicals, or reactants.

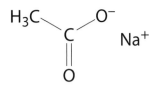

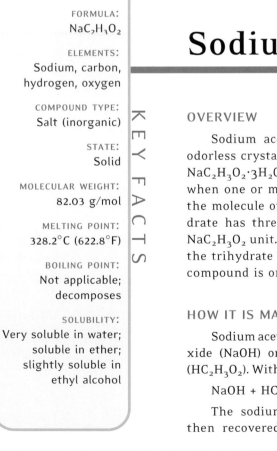

Sodium Acetate

OVERVIEW

Sodium acetate (SO-dee-um ASS-uh-tate) is a colorless, odorless crystalline solid that often occurs as the trihydrate: $NaC_2H_3O_2 \cdot 3H_2O$. A hydrate is a chemical compound formed when one or more molecules of water is physically added to the molecule of some other substance. Sodium acetate trihydrate has three molecules of water of hydration for every $NaC_2H_3O_2$ unit. Anhydrous sodium acetate readily converts to the trihydrate because it is very hygroscopic. A hygroscopic compound is one that readily absorbs moisture from the air.

HOW IT IS MADE

Sodium acetate is prepared by reacting either sodium hydroxide (NaOH) or sodium carbonate (Na_2CO_3) with acetic acid ($HC_2H_3O_2$). With sodium hydroxide, for example, the reaction is:

$$NaOH + HC_2H_3O_2 \rightarrow NaC_2H_3O_2 + H_2O$$

The sodium acetate forms as the trihydrate, which is then recovered by evaporating the reacting solution. The

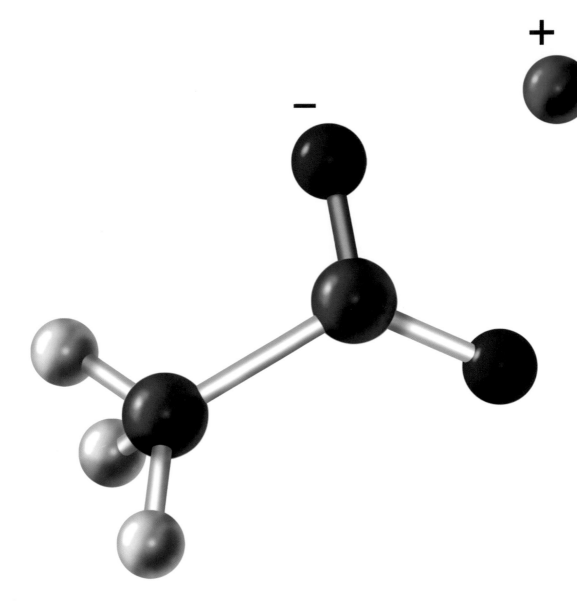

Sodium acetate trihydrate.
Red atoms are oxygen;
hydrogen atoms are white;
black atoms are carbon;
turquoise atom is sodium. Gray
stick indicates double bond.

anhydrous salt can be prepared by heating the hydrate to drive off the water of hydration.

COMMON USES AND POTENTIAL HAZARDS

Sodium acetate is used in a number of industries. In the textile industry, it is used as a mordant in the dyeing of fabrics and to stabilize and improve the finish of fabrics. The

Interesting Facts

- In 1995, officials in Michigan tested sodium acetate trihydrate as a deicing salt for airport runways. Experiments showed, however, that the anhydrous salt worked better than the trihydrate for this purpose.

- Sodium acetate has also been used as a deicer in parking garages. The compound is preferred to sodium chloride, widely used as a deicing compound, because sodium chloride corrodes steel rods buried in concrete and sodium acetate does not.

- Heat packs sometimes contain a solution of sodium acetate trihydrate cooled below its freezing point. When the pack is activated, the compound freezes very rapidly, releasing the heat that had been stored in its super-cooled phase.

cosmetics industry uses sodium acetate as a buffering agent in a variety of personal care products. A buffering agent is a substance that maintains the acidity of a product within a certain desired range. Sodium acetate is also used by food producers for the same reason, assuring that a variety of foods have an acidity sufficient to protect the food from decaying, but not so acidic as to have an unpleasant taste. Some other applications of sodium acetate include:

- In heat packs to relieve stiffness and pain, to keep hands and feet warm, and to warm baby bottles;

- In the production of soaps, where it reacts with strong bases to reduce the harshness of the final product;

- In dialysis machines, used for people whose kidneys are not working properly, to provide the sodium ions (Na^+) to maintain proper electrolyte balance in the body;

- As a diuretic, a drug used to promote urination;

- As an expectorant in drugs used to promote coughing to help bring up mucous;

Words to Know

ANHYDROUS Lacking water of hydration.

MORDANT A substance used in dyeing and printing that reacts chemically with both a dye and the material being dyed to help hold the dye permanently to the material.

- As a veterinary treatment for bovine ketosis, a condition caused by low blood sugar in cows that results in a wasting or weakening of the animal;

- As a buffer in the developing of photographs;

- In the tanning of hides to obtain a more even and more rapid absorption of the tanning material; and

- In the purification of glucose.

Sodium acetate is a mild irritant to the skin, eyes, and respiratory system. If inhaled, it may cause inflammation of the throat and lungs. At the level it appears in most household products, it presents a very low hazard to the average person.

FOR FURTHER INFORMATION

"How Do Sodium-Acetate Heat Pads Work?" How Stuff Works. http://home.howstuffworks.com/question290.htm (accessed on November 5, 2005).

"Material Safety Data Sheet: Sodium Acetate Trihydrate." Iowa State University, Department of Chemistry. http://avogadro.chem.iastate.edu/MSDS/NaOAc-3H2O.htm (accessed on November 5, 2005).

"Sodium Acetate, Anhydrous." Cornell Material Safety Data Sheets. http://msds.ehs.cornell.edu/msds/msdsdod/a73/m36008.htm#Section3 (accessed on November 5, 2005).

"Sodium Acetate Anhydrous—Physical Properties." Jarchem Industries. http://www.jarchem.com/sodium-acetate-anhydrous.htm (accessed on November 5, 2005).

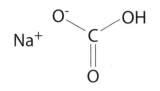

KEY FACTS

Sodium Bicarbonate

OVERVIEW

Sodium bicarbonate (SO-dee-um bye-KAR-bun-ate) is a white, odorless, crystalline solid or powder that is stable in dry air, but that slowly decomposes in moist air to form sodium carbonate. The compound's primary uses are as an additive in human and animal food products.

Sodium bicarbonate has been used by humans for thousands of years. Ancient Egyptian documents mention the use of a sodium bicarbonate and sodium chloride solution in the mummification of the dead. For centuries, people around the world have used sodium bicarbonate as a leavening agent for baking. A leavening agent is a substance that causes dough or batter to rise. Sodium bicarbonate produces this effect because, when heated or dissolved in water, it breaks down to produce carbon dioxide (CO_2) gas:

$$2NaHCO_3 \rightarrow Na_2CO_3 + CO_2 + H_2O$$

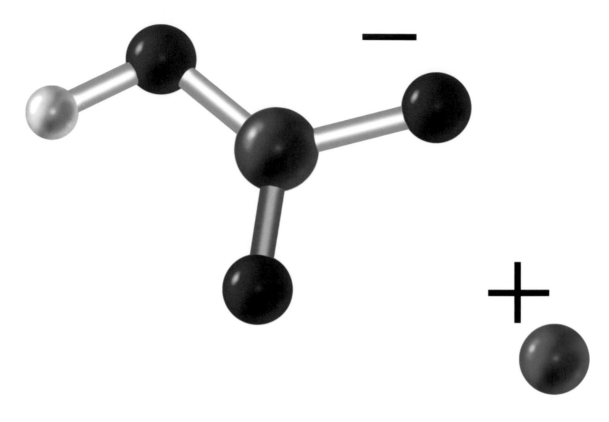

Sodium bicarbonate. Red atoms are oxygen; white atom is hydrogen; black atom is carbon; and turquoise atom is sodium. Gray stick indicates double bond. PUBLISHERS RESOURCE GROUP

Since all the compounds present in this reaction are safe for human consumption, sodium bicarbonate makes an ideal leavening agent.

Commercial production of sodium bicarbonate as baking soda dates to the late 1700s. In 1846, Connecticut physician Austin Church (1799-?) and John Dwight (1819-?) of Dedham, Massachusetts, founded a company to make and sell sodium bicarbonate. They started their company in the kitchen of Dwight's home, making the product by hand and packing it in paper bags for sale to neighbors. The Church-Dwight operation grew over the years to become the largest producer of household baking soda, now sold under the name of Arm & Hammer® baking soda. The company still produces 90 percent of all the baking soda used for household purposes in the United States. Consumers use the product for cooking, cleaning, and deodorizing homes.

Interesting Facts

- Sodium bicarbonate is a very effective cleaning agent for certain materials. In the 1980s, restorers used an aqueous solution of the compound to clean the Statue of Liberty.

HOW IT IS MADE

Sodium bicarbonate is made commercially by one of two methods. In the first method, carbon dioxide gas is passed through an aqueous solution of sodium carbonate (Na_2CO_3):

$$Na_2CO_3 + CO_2 + H_2O \rightarrow 2NaHCO_3$$

Since the bicarbonate is less soluble than the carbonate, it precipitates out of solution and can be removed by filtration.

Sodium bicarbonate is also obtained as a byproduct of the Solvay process. The Solvay process was invented in the late 1850s by Belgian chemist Ernest Solvay (1838–1922) primarily as a way of making sodium carbonate. Sodium carbonate had long been a very important industrial chemical for which no relatively inexpensive method of preparation existed. Solvay developed a procedure by which sodium chloride is treated with carbon dioxide and ammonia, resulting in the formation of sodium bicarbonate and ammonium bicarbonate. The sodium bicarbonate is then heated to obtain sodium carbonate. Although sodium carbonate is the desired product in this reaction, sodium bicarbonate can also be obtained by deleting the final step by which it is converted into sodium carbonate.

COMMON USES AND POTENTIAL HAZARDS

An estimated 560,000 metric tons (615,000 short tons) of sodium bicarbonate were consumed in the United States in 2003. About one-third of that amount was used by the food products industry, primarily in the manufacture of baking soda (pure sodium bicarbonate) and baking powder (a mixture of sodium bicarbonate and at least one other compound).

Baking powder differs from baking soda in that it includes an acidic compound that reacts with sodium bicarbonate to produce carbon dioxide. One of the most common compounds mixed with sodium bicarbonate in baking powder is tartaric acid ($HOOC(CHOH)_2COOH$), or its salt, potassium bitartrate ($HOOC(CHOH)_2COOK$). Baking powder is a more efficient leavening agent in baking than is baking soda by itself. Baking soda is also used as an additive in foods and drinks to provide effervescence (a bubbling, fizzing, or sparkling effect) or to maintain an acidic environment in the food. The acidity provides a sharp taste and helps to preserve a food.

The second largest use of sodium bicarbonate is as an additive in animal feed. As with human foods, it maintains the proper acidity of an animal's feed, improving its ability to digest and absorb its food.

Sodium bicarbonate is also used in a number of pharmaceutical applications. For example, it is a common ingredient in antacids, products designed to relieve heartburn, acid indigestion, sour stomach, and other discomforts caused by overeating or improper foods. Some pharmaceuticals, such as Alka-Seltzer®, contain a combination of citric acid and sodium bicarbonate. The citric acid helps the sodium bicarbonate dissolve more quickly and produces more effervescence when the tablet is dissolved in water.

Sodium bicarbonate is also used in cleaning products on both a household and industrial level. Many householders use commercial baking soda, such as that sold by the Arm & Hammer company, to clean kitchen and bathroom appliances, such as sinks, stoves, and toilet bowls. Industries also use sodium bicarbonate filters to remove sulfur dioxide and other pollutants in flu gases released from factory smokestacks. The compound is also used in the treatment of wastewater to maintain proper acidity, remove certain odors (such as those of sulfur dioxide), and destroy bacteria. Some communities have used aqueous solutions of sodium bicarbonate sprayed at high pressure to remove graffiti; paint; soot and smoke residues; and mold from buildings, walls, and other public structures.

Some other applications of sodium bicarbonate include:

- As a component of fire extinguishers; when it comes into contact with an acid in the fire extinguisher, the sodium bicarbonate releases carbon dioxide and a flow of water under pressure to fight the fire;

Words to Know

AQUEOUS SOLUTION A solution that consists of some material dissolved in water.

PRECIPITATE A solid material that settles out of a solution, often as the result of a chemical reaction.

- As a blowing agent in the preparation of plastics; blowing agents are substances that produce large volumes of gas that convert a molten product into a foamy product;
- In the manufacture of other sodium compounds;
- For gold and platinum plating; and
- To prevent the growth of mold on timber.

Sodium bicarbonate is considered safe when handled or ingested in reasonable amounts. As with any chemical, however, excessive amounts of the compound can have harmful effects. When ingested in large amounts, sodium bicarbonate can cause stomach cramps, gas, upset stomach, vomiting, frequent urination, loss of appetite, and blood in the urine and stools.

FOR FURTHER INFORMATION

"Pure Baking Soda." Arm & Hammer®.
http://www.armhammer.com/ (accessed on November 8, 2005).

Snyder, C. H. *The Extraordinary Chemistry of Ordinary Things*, 4th ed. New York: John Wiley and Sons, 2002.

"Sodium Bicarbonate." Chemical Land 21.
http://www.chemicalland21.com/arokorhi/industrialchem/inorganic/SODIUM%20BICARBONATE.htm (accessed on November 8, 2005).

"Sodium Bicarbonate." DC Chemical Co., Ltd.
http://www.dcchem.co.kr/english/product/p_basic/p_basic02.htm (accessed on November 8, 2005).

See Also Carbon Dioxide; Citric Acid; Sodium Carbonate

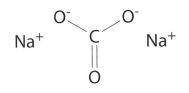

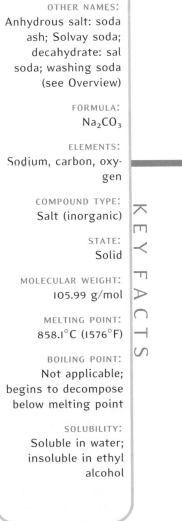

OTHER NAMES:
Anhydrous salt: soda
ash; Solvay soda;
decahydrate: sal
soda; washing soda
(see Overview)

FORMULA:
Na_2CO_3

ELEMENTS:
Sodium, carbon, oxy-
gen

COMPOUND TYPE:
Salt (inorganic)

STATE:
Solid

MOLECULAR WEIGHT:
105.99 g/mol

MELTING POINT:
858.1°C (1576°F)

BOILING POINT:
Not applicable;
begins to decompose
below melting point

SOLUBILITY:
Soluble in water;
insoluble in ethyl
alcohol

KEY FACTS

Sodium Carbonate

OVERVIEW

Sodium carbonate (SO-dee-um KAR-bun-ate) is an odorless white powder or crystalline solid with an alkaline taste. (Baking soda is another substance with an alkaline taste.) It is hygroscopic, meaning that it has a tendency to absorb moisture from the air. It also exists as the monohydrate ($Na_2CO_3 \cdot H_2O$) and as the decahydrate ($Na_2CO_3 \cdot 10H_2O$), each with slightly different physical properties from those of the anhydrous salt. The anhydrous form of sodium carbonate is commonly known as soda ash, while the decahydrate is often called sal soda or washing soda. Sodium carbonate has long been one of the most important chemical compounds produced in the United States. Its primary use is in the manufacture of glass and other chemicals.

HOW IT IS MADE

Humans have known about and used sodium carbonate for thousands of years. The ancient Egyptians extracted the compound from a mineral known as natron found in dry lake

Interesting Facts

- Over the last decade, consumption of sodium carbonate in the United States has been decreasing, largely because glass containers (made with sodium carbonate) are being replaced by plastic containers. However, the industry has not gone into decline primarily because the demand for glass in developing countries continues to increase at a rate that matches the decline in glass bottle demand in the United States.

on the type of glass one wishes to make. For example, oxides of various metals such as iron(III) oxide and copper(II) oxide are added to provide a reddish tint to the glass.

The second highest use of sodium carbonate is in the production of other chemical compounds, followed by the compound's use in the production of soaps and detergents. Other applications of the compound include:

- The production of pulp and paper products;
- The removal of sulfur dioxide from flu gases in factories;
- In water purification and waste water treatment facilities;
- As a mordant in the dyeing of cloth;
- In the refining of petroleum;
- As a catalyst in the process by which coal is converted into a liquid fuel;
- For the bleaching of cotton and linen fabrics;
- As an emetic (a compound that induces vomiting); and
- In cosmetics and personal care products because of its ability to clean skin and help clear up rashes.

No serious health hazards have been associated with the use of sodium carbonate in any of its forms.

Words to Know

ANHYDROUS Lacking water of hydration.

CATALYST A material that increases the rate of a chemical reaction without undergoing any change in its own chemical structure.

MORDANT A substance used in dyeing and printing that reacts chemically with both a dye and the material being dyed to help hold the dye permanently to the material.

SYNTHESIS A chemical reaction in which some desired chemical product is made from simple beginning chemicals, or reactants.

WATER OF HYDRATION Water that has combined with a compound by some physical means.

FOR FURTHER INFORMATION

Kiefer, David M. "Soda Ash, Solvay Style." *Today's Chemist* (February 2002): 87-88+. Also available online at http://pubs.acs.org/subscribe/journals/tcaw/11/i02/html/02chemchron.html (accessed on November 8, 2005).

Lister, Ted, compiler. "Sodium Carbonate—A Versatile Material." Royal Society of Chemistry. http://www.chemsoc.org/pdf/LearnNet/rsc/SodiumCarb_-sel.pdf (accessed on November 8, 2005).

Monet, Jefferson. "An Overview of Mummification in Ancient Egypt." TourEgypt.net. http://www.touregypt.net/featurestories/mummification.htm (accessed on November 8, 2005).

"Soda Ash or Trona." Mineral Information Institute. http://www.mii.org/Minerals/phototrona.html (accessed on November 8, 2005).

"Sodium Carbonate." United Nations Environmental Programme. http://www.inchem.org/documents/sids/sids/Naco.pdf (accessed on November 8, 2005).

See Also Sodium Bicarbonate

Interesting Facts

- Salt was a key ingredient in the solution used by ancient Egyptians in the preparation of mummies.

- The expression "not worth his salt" had its origin in the ancient Greek slave trade, in which people were bought and sold for measures of salt.

- The English word *salary* comes from the Latin term *salarium argentums*, which refers to a special salt ration given to Roman soldiers.

- Salt has frequently been subject to heavy taxation; the very high tax on salt in France during the middle eighteenth century contributed to the rise of the French revolution.

- Salt mines that are no longer in use are sometimes used to store natural gas and petroleum.

- As a feed additive for livestock, poultry, and other domestic animals, to ensure that they receive the sodium and chlorine they need to remain healthy and grow normally;

- As a deicing product on roads and highways;

- In the manufacture of glazes used on ceramic products;

- For the curing of animal hides;

- In the dyeing and printing of fabrics;

- In the manufacture of soaps;

- As a herbicide, a chemical used to kill weeds; and

- As a fire extinguisher for certain types of fires (such as grease fires).

Given its widespread use, sodium chloride is obviously safe for consumption by most humans under normal conditions. As with any chemical compound, consumption of a large excess of sodium chloride can be harmful. The one health issue of greatest concern has to do with high blood pressure. Scientists have learned that the ingestion of large amounts of sodium can contribute to hypertension (high blood pressure), which in turn is associated with increased risk for heart attacks and stroke.

Words to Know

PHARMACOLOGY The study of compounds used as drugs.

The American Heart Association recommends that healthy American adults consume no more than 2,300 milligrams of sodium a day. That amount is equivalent to about a teaspoon of salt. For those considered at higher risk, people with high blood pressure, blacks, and middle-aged and older adults, the 2005 U.S. Department of Agriculture guidelines recommend no more than 1,500 milligrams of sodium per day. The problem with sodium chloride consumption is that most people have no idea how much salt they eat every day. Of course, they can keep track of the salt they add to the foods they prepare in their own homes. But most commercially prepared foods also have sodium chloride added to them. In some cases, the total amount of salt ingested from processed foods by the average American can be significant, easily exceeding the recommended daily average recommended by the American Heart Association. People can, therefore, be consuming dangerously high levels of sodium without being aware of that fact.

FOR FURTHER INFORMATION

"History of Salt." The Salt Institute.
http://www.saltinstitute.org/38.html (accessed on November 8, 2005).

Kurlansky, Mark. *Salt: A World History.* New York: Walker, 2002.

"Salt." History for Kids.
http://www.historyforkids.org/learn/food/salt.htm (accessed on November 8, 2005).

"Sodium Chloride." J. T. Baker.
http://www.jtbaker.com/msds/englishhtml/S3338.htm (accessed on November 8, 2005).

"What You Always Wanted to Know about Salt." The Salt Institute.
http://www.saltinstitute.org/4.html (accessed on November 8, 2005).

See Also Chlorine; Sodium Hydroxide

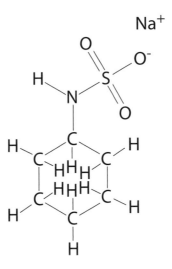

Sodium Cyclamate

OVERVIEW

Sodium cyclamate (SO-dee-um SYE-kla-mate) is a white, crystal solid or powder with almost no odor and a very sweet taste. Its sweetening power is about 30 times that of table sugar, the standard against which artificial sweeteners are measured. Because of its sweet flavor, sodium cyclamate is used as an artificial sweetener.

The cyclamate family of compounds was discovered in 1937 by Michael Sveda (1912-1999), then a graduate student at the University of Illinois. Sveda was working on the development of new drugs to treat fever. The story is that Sveda was smoking while he was working in the laboratory (a practice that would not be allowed today) and, at one point, he brushed some loose threads of tobacco from his lips. As he did so, he noticed a very sweet flavor on the cigarette. Curious as to the cause of the sweetness, Sveda looked more carefully into the compounds he was studying and eventually identified the source of the sweetness as a substance belonging to a class of compounds known as cyclamates. The cyclamates

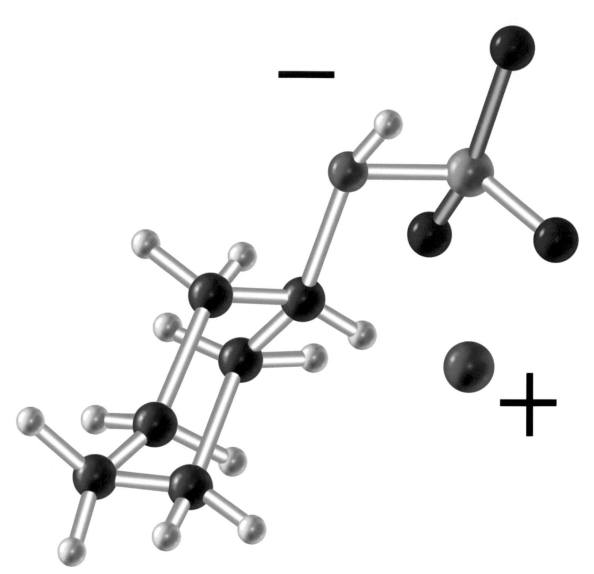

Sodium cyclamate. Red atoms are oxygen; white atoms are hydrogen; black atoms are carbon; blue atom is nitrogen; yellow atom is sulfur; and turquoise atom is sodium. Gray sticks indicate double bonds.

are organic salts of the carboxylic acid cyclamic acid ($C_6H_{11}NHSO_3H$). The University of Illinois received a patent for the production of cyclamates, which they eventually sold to the DuPont chemical company. DuPont, in turn, sold the patent to Abbott Laboratories, who first marketed the product in 1950. The compound rapidly became very popular as an additive for diet foods and drinks. It most formulations, it is mixed with saccharin, another artificial sweetener. Saccharin is ten times as sweet as sodium cyclamate, but it leaves a metallic aftertaste that sodium cyclamate helps to mask.

The first evidence about possible health effects of ingesting cyclamates was announced in 1969. Scientists reported that they had fed eighty rats a diet that contained large amounts of Sucaryl®, a product containing saccharin and sodium cyclamate, every day for two years. At the end of that time, twelve of the eighty rats had developed bladder cancer. Based on this study, the U.S. Food and Drug Administration (FDA) decided to ban the use of cyclamates as food additives.

The FDA's decision has long been the subject of considerable debate. Some people praise the agency for banning a substance they believe to be carcinogenic in humans as well as rats. Others feel that the evidence on which the FDA decision was based is not strong enough to require banning of the product. In light of this controversy, a number of additional studies have been conducted on the health effects of the cyclamates. In 1984, the FDA reversed its decision on the compound, having become convinced that no evidence had been produced to support the original 1969 study on which its earlier decision had been made. A year later, the National Academy of Sciences announced the results of its own review of the evidence on the cyclamates. The academy also saw no reason to ban the product as a food additive. Outside the United States, the cyclamates have generally been approved for use as an artificial sweetener.

HOW IT IS MADE

Cyclamic acid is made by the sulfonation of cyclohexylamine. Cyclohexylamine is a six-carbon ring compound with a single amine (-NH_2) group attached to it. Its formula is $C_6H_{11}NH_2$. Sulfonation is the process by which an -SO_2 group is added to a compound. Sulfonation of cyclohexylamine is accomplished with either sulfur dioxide (SO_2) or sulfamic acid ($HOSO_2NH_2$).

COMMON USES AND POTENTIAL HAZARDS

Sodium cyclamate and calcium cyclamate are used as artificial sweeteners. Since they have no nutritional value

Interesting Facts

- Sodium cyclamate was nearly the undoing of the sports drink Gatorade. The drink was invented in the late 1960s. The inventors decided to use sodium cyclamate as a sweetener because it had less of a bitter aftertaste than saccharin. One year after Gatorade first appeared, however, the FDA banned cyclamates as food additives. The inventors quickly reformulated their product, using fructose in place of cyclamates as a sweetener. The drink went on to become an outstanding commercial success, but without the benefit of sodium cyclamate.

(that is, the contain no calories), they can be used in foods and drinks designed for diabetics and dieters. The product also appeals to food processors because it is much sweeter than table sugar. One gram of sodium or calcium cyclamate is as sweet as 30 grams of table sugar, so much less is needed to make a product taste sweet. The cyclamates are also stable to heat, which means that they will not break down if foods in which they are contained are baked or boiled. The compounds have long shelf lives and are inexpensive to make.

Cyclamates are often used in combination with table sugar and other artificial sweeteners. The combination of a cyclamate and saccharin, for example, has the benefit that the two sweeteners cancel out the bitter and metallic aftertastes that each by itself has.

Words to Know

CARCINOGEN A substance that causes cancer in humans or other animals.

FOR FURTHER INFORMATION

"Cyclamate." Zhonga Hua Fang Da.
http://www.fangda.com.hk/english/ (accessed on November 8, 2005).

"Low-Calorie Sweeteners: Cyclamate." CalorieControl.org.
http://www.caloriecontrol.org/cyclamat.html (accessed on November 8, 2005).

"Sodium Cyclamate." Hazard Database.
http://www.evol.nw.ru/labs/lab38/spirov/hazard/sodium_cyclamate.html (accessed on November 8, 2005).

See Also L-Aspartyl-L-phenylalanine Methyl Ester; Saccharin; Sucrose

F⁻   Na⁺

OTHER NAMES:
Sodium monofluoride

FORMULA:
NaF

ELEMENTS:
Sodium, fluorine

COMPOUND TYPE:
Binary salt (inorganic)

STATE:
Solid

MOLECULAR WEIGHT:
41.99 g/mol

MELTING POINT:
996°C (1824°F)

BOILING POINT:
1704°C (3099°F)

SOLUBILITY:
Moderately soluble in water; insoluble in ethyl alcohol

KEY FACTS

Sodium Fluoride

OVERVIEW

Sodium fluoride (SO-dee-um FLOR-ide) is a colorless to white crystalline solid or powder. It is best known for its role in efforts to prevent tooth decay. It may be added to toothpastes or mouthwashes or to municipal water supplies for this purpose. Although the practice of fluoridating water is now widespread in the United States, it remains the subject of controversy regarding its potential health effects on humans.

HOW IT IS MADE

Sodium fluoride occurs naturally as the mineral villiaumite, although the compound is not produced commercially from that source. Some sodium fluoride is obtained as a byproduct of the manufacture of phosphate fertilizers. In that process, apatite (a form of calcium phosphate that also contains fluorides and/or chlorides) is crushed and treated with sulfuric acid (H_2SO_4). The products of that reaction include phosphoric acid (H_3PO_4), calcium sulfate ($CaSO_4$),

Sodium fluoride. Green atom is fluorine and turquoise atom is sodium. PUBLISHERS RESOURCE GROUP

hydrogen fluoride (HF), and silicon tetrafluoride (SiF_4). The hydrogen fluoride and silicon tetrafluoride can then be converted into sodium fluoride. The compound can also be produced by treating hydrogen fluoride with sodium carbonate (Na_2CO_3):

$$2HF + Na_2CO_3 \rightarrow 2NaF + H_2O + CO_2$$

COMMON USES AND POTENTIAL HAZARDS

More than 50 years of research has shown that sodium fluoride and other fluorides are effective in preventing tooth decay. Based on this information, sodium fluoride or some other compound of fluorine is now added to most toothpastes made in the United States. Dentists regularly treat their patients' teeth with fluoride washes to make them more resistant to decay. Most cities and towns in the United States add sodium fluoride or a comparable compound to the municipal water supply to reduce the rate of dental caries (cavities). Some people who live where fluoride is not added to their water supply take sodium fluoride pills to improve their dental health. The American Dental Association and the World Health Organization recommend

Interesting Facts

- Some of the opposition to the fluoridation of water is based on a misunderstanding of the difference between fluorine, the element, and fluorides, compounds of fluorine, such as sodium fluoride and potassium fluoride. Fluorides have very different physical, chemical, and biological properties from the element fluorine. For example, fluorine gas is a highly toxic gas that reacts violently with most substances, including water. Flourides, on the other hand, are relatively inert (unreactive) and safe to ingest in small amounts.

fluoridation of drinking water at a level between 0.7 and 1.2 parts per million.

In spite of these trends, opposition to the use of fluorides against tooth decay remains strong in the United States and other parts of the world. Opponents are not convinced that there is sufficient evidence for the claims that fluoridation decreases the rate of tooth decay. They suggest that fluorides may cause cancer and a host of other health problems. And they argue that fluoridating public water supplies removes the choice that individuals should have as to whether or not they want to use fluorides in their dental health program.

Sodium fluoride has a number of commercial and industrial uses in addition to those related to dental health. Those uses include:

- As a wood preservative;

- In the production of certain types of pesticides and as an insecticide for ant and roach control;

- In the preparation of other fluoride salts;

- In electroplating operations;

Words to Know

ELECTROPLATING A process by which a thin layer of one metal is deposited on top of a second metal by passing an electric current through a solution of the first metal.

- In the degassing (removal of gas pockets) during the manufacture of steel;

- For the manufacture of glass and vitreous (glass-like) enamels;

- In detection systems for radiation in the ultraviolet and infrared regions of the electromagnetic spectrum; and

- For disinfecting equipment used in breweries and wineries.

The discussion about fluoridation of public water supplies is sometimes made more difficult because of the fact that sodium fluoride poses some real health hazards to humans. It is a skin, eye, and respiratory irritant that can cause burning if spilled on the skin or in the eyes. If swallowed, it can cause burning of the digestive tract, nausea, vomiting, abdominal pain, stupor, general weakness, tremors, convulsions, collapse, respiratory and cardiac failure, and death. Ingestion of as little as five grams of sodium fluoride can result in death. A condition known as fluorosis is also associated with high doses of sodium fluoride. Fluorosis is characterized by a yellowing and increased brittleness of teeth and bones. Although these conditions are all very serious, they appear only at doses many thousands of times greater than one receives from fluorides in toothpastes and public water supplies.

FOR FURTHER INFORMATION

"Facts about Fluoride." American Dental Association. http://www.ada.org/public/topics/fluoride/fluoride_article01.asp (accessed on November 8, 2005).

"Fluoride in Drinking Water." *Science in Dispute.* Ed. Neil Schlager. Vol. 1. Detroit: Gale, 2002.

"Sodium Fluoride." Fluoride Action Network. http://www.flouridealert.com/pesticides/sodium-fluoride-page .htm (accessed on November 8, 2005).

"Sodium Fluoride." J. T. Baker. http://www.jtbaker.com/msds/englishhtml/S3722.htm (accessed on November 8, 2005).

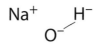

Na$^+$ H$^-$
    O$^-$

OTHER NAMES:
Caustic soda; lye;
sodium hydrate;
white caustic

FORMULA:
NaOH

ELEMENTS:
Sodium, oxygen,
hydrogen

COMPOUND TYPE:
Base (inorganic)

STATE:
Solid

MOLECULAR WEIGHT:
40.00 g/mol

MELTING POINT:
323°C (613°F)

BOILING POINT:
1388°C (2530°F)

SOLUBILITY:
Soluble in water,
ethyl alcohol, methyl
alcohol, and glycerol

Sodium Hydroxide

OVERVIEW

Sodium hydroxide (SO-dee-um hye-DROK-side) is a white deliquescent solid commercially available as sticks, pellets, lumps, chips, or flakes. A deliquescent material is one that absorbs moisture from the air. Sodium hydroxide also reacts readily with carbon dioxide in the air to form sodium carbonate. Sodium hydroxide is the most important commercial caustic. A caustic material is a strongly basic or alkaline material that irritates or corrodes living tissue. The compound ranked number 11 among chemicals produced in the United States in 2004.

HOW IT IS MADE

Sodium hydroxide is produced commercially simultaneously with chlorine gas by the electrolysis of a sodium chloride solution. In this process, an electric current breaks down sodium chloride into its component elements, sodium and chlorine. The chlorine escapes as a gas, while the sodium metal form reacts with water to form sodium hydroxide:

$$2NaCl \rightarrow 2Na + Cl_2$$

$$2Na + 2H_2O \rightarrow 2NaOH + H_2$$

Sodium hydroxide can also be produced easily by means of other chemical reactions. For example, the reaction between slaked lime (calcium hydroxide; $Ca(OH)_2$) and soda ash (sodium carbonate; Na_2CO_3) produces sodium hydroxide:

$$Ca(OH)_2 + Na_2CO_3 \rightarrow 2NaOH + CaCO_3$$

None of these alternative methods can compete economically, however, with the preparation by electrolysis.

Interesting Facts

- Solutions of sodium hydroxide are made by adding the solid compound to water, and never water to the solid. The reason is that large amounts of heat are generated when sodium hydroxide dissolves in water. That heat is absorbed by water, but would not be absorbed by solid sodium hydroxide.

- A popular food in Scandinavian countries, lutefisk, is made by soaking dried fish in sodium hydroxide until it turns into a jelly. The jelly is then soaked in water for several days to remove the poisonous lye.

COMMON USES AND POTENTIAL HAZARDS

Sodium hydroxide has a great variety of household and industrial uses. It is the active ingredient in drain cleaners such as Drano® because it breaks up and dissolves the greasy mass that is responsible for drain blockages. It is also an ingredient in many other household products, including oven cleaners, metal polishes, and hair straighteners. Sodium hydroxide is also used in the preparation of homemade and processed foods. It is used in the preparation of soft drinks, chocolate, ice creams, caramel coloring, and cocoa. Hominy, a starchy food similar to grits, is made by soaking corn kernels in a solution of sodium hydroxide in water. Bakers glaze pretzels and German lye rolls with a weak lye solution before baking them. The lye gives baked goods a crisp crust. Some people use lye to cure olives.

The largest single use for sodium hydroxide is in the production of organic compounds from which polymers are made, such as propylene oxide and the ethylene amines, and of the polymers themselves, including the polycarbonates and epoxy resins. About a third of all the sodium hydroxide produced in the United States goes to this application. Another important use of sodium hydroxide is in the pulp and paper industry, where it is used to digest (break down) the raw materials from which pulp and paper are made. About 13 percent of all the sodium hydroxide made in the

Words to Know

ELECTROLYSIS A process in which an electric current is used to bring about chemical changes.

HYDROLYSIS The process by which a compound reacts with water to form two new compounds.

United States goes to this application. Sodium hydroxide is also an important raw material in the manufacture of soap. The method by which soap is made has not changed very much for thousands of years. A fat or oil is added to a boiling solution of sodium hydroxide in water. The fat or oil hydrolyzes into its component parts, glycerol and fatty acids. The sodium hydroxide then reacts with the fatty acids, forming sodium salts. The sodium salt of a fatty acid is a soap. Sodium hydroxide is also an important raw material in the manufacture of inorganic compounds, especially sodium and calcium hypochlorite, sodium cyanide, and a number of sulfur-containing compounds. Some other important uses of sodium hydroxide include:

- In the manufacture of cellophane and rayon;
- As a neutralizing agent during the refining of petroleum;
- In the manufacture of aluminum metal;
- For the refining of vegetable oils;
- As an agent for peeling fruits and vegetables for processing;
- In the extraction of metals from their ores;
- For the processing of textiles;
- In water treatment facilities;
- For etching and electroplating operations; and
- In a wide variety of research laboratory applications.

Sodium hydroxide is one of the most caustic substances known and a strong irritant to the skin, eyes, and respiratory system. Exposure to sodium hydroxide dust, powder, or solid can cause burning of the skin and eyes, with possible permanent damage to one's vision. Ingestion of the compound causes burning of the mouth, esophagus, and stomach, resulting in nausea, diarrhea, internal bleeding, scarring, and permanent damage to the lungs and gastrointestinal system. More serious results, such as a drop in blood pressure and collapse, are also possible.

FOR FURTHER INFORMATION

"Determination of Acute Reference Exposure Levels for Airborne Toxicants." [Sodium Hydroxide]. Office of Environmental Health Hazard Assessment, State of California. http://www.oehha.ca.gov/air/acute_rels/pdf/1310932A.pdf (accessed on November 8, 2005).

"DOW Caustic Soda Solution." Dow Chemical Company. http://www.dow.com/causticsoda/prod/process.htm (accessed on November 8, 2005).

"Sodium Hydroxide." International Chemical Safety Cards. http://www.cdc.gov/niosh/ipcsneng/neng0360.html (accessed on November 8, 2005).

"Sodium Hydroxide." Medline Plus. http://www.nlm.nih.gov/medlineplus/ency/article/002487.htm (accessed on November 8, 2005).

White, Elaine. "Making Modern Soap with Herbs, Beeswax, and Vegetable Oils." http://www.pioneerthinking.com/soaps.html (accessed on November 8, 2005).

See Also Chlorine; Potassium Hydroxide

$$Na^+ \quad Cl^-$$
$$\diagdown \diagup$$
$$O$$

OTHER NAMES:
Sodium oxychloride;
hypochlorite; bleach;
chlorine bleach

FORMULA:
NaClO

ELEMENTS:
Sodium, chlorine,
oxygen

COMPOUND TYPE:
Oxy salt (inorganic)

STATE:
Solid or aqueous
solution; See Over-
view

MOLECULAR WEIGHT:
74.44 g/mol

MELTING POINT:
Solid NaClO explodes
on heating.
The pentahydrate
(NaClO·5H₂O) is more
stable; its melting
point is 18°C (64°F)

BOILING POINT:
Not applicable;
decomposes

SOLUBILITY:
Soluble in water

KEY FACTS

Sodium Hypochlorite

OVERVIEW

Sodium hypochlorite (SO-dee-um hye-po-KLOR-ite) is the active ingredient in liquid chlorine bleaches, used in the home and many industries to whiten fabric and other materials and to disinfect surfaces and water. The anhydrous compound is very unstable and explodes readily. The pentahydrate is a pale-green crystalline solid that is relatively stable. The compound is usually made available as an aqueous solution that contains anywhere from 3 to 6 percent sodium hypochlorite (for house-hold use) to as high as 30 percent (for industrial applications). In solution form, sodium hypochlorite is quite stable and can be stored for long periods of time out of sunlight.

Sodium hypochlorite decomposes by two mechanisms. In one case, it breaks down to form sodium chloride and sodium chlorate:

$$3NaOCl \rightarrow 2NaCl + NaClO_3$$

In the second case, it breaks down to form sodium chloride and nascent (free single atoms) oxygen:

Interesting Facts

- Until the discovery of chlorine, cloth was usually bleached by soaking it in sour milk or buttermilk and letting it sit in the sun. The process often took up to eight weeks and required large "bleaching fields" on which the cloth could be laid out.

- The first attempt to apply chemical principles to the practice of bleaching was documented in a book on the subject by the Scottish physician Francis Home, published in 1756. Home suggested using a weak solution of sulfuric acid for bleaching, a practice that reduced bleaching time to about 12 hours.

mable mixture. The fumes from this combination can also be harmful, even deadly. Similarly, household bleach should not be use to clean spills that contain urine since urine itself contains ammonia. Sodium hypochlorite is also incompatible with hydrogen peroxide and acidic products.

Sodium hypochlorite is an irritant to the skin, eyes, and respiratory system. It can produce inflammation, burning, and blistering of the skin; burning of the eyes, with subsequent damage to one's vision; and irritation of the gastrointestinal system that can result in stomach pain, nausea, vomiting, coughing, and ulceration of the digestive tract.

Words to Know

ANHYDROUS Lacking water of hydration.

AQUEOUS SOLUTION A solution that consists of some material dissolved in water.

OXIDATION A chemical reaction in which oxygen reacts with some other substance or, alternatively, in which some substance loses electrons to another substance, the oxidizing agent.

PENTAHYDRATE A form of a crystalline compound that occurs with five molecules of water.

CHEMICAL COMPOUNDS

FOR FURTHER INFORMATION

Chalmers, Louis. *Household and Industrial Chemical Specialties.* Vol. 1. New York: Chemical Publishing Co., Inc., 1978.

Fletcher, John, and Don Ciancone. "Why Life's a Bleach (The Sodium Hypochlorite Story)." *Environmental Science & Engineering.* May 1996. Also available online at http://www.esemag.com/0596/bleach.html (accessed on November 8, 2005).

"Medical Management Guidelines (MMGs) for Calcium Hypochlorite ($CaCl_2O_2$) Sodium Hypochlorite (NaOCl)." Agency for Toxic Substances and Disease Registry. http://www.atsdr.cdc.gov/MHMI/mmg184.html (accessed on January 12, 2006).

"Sodium Hypochlorite." Hill Brothers Chemical Co. http://hillbrothers.com/msds/pdf/sodium-hypochlorite.pdf (accessed on November 8, 2005).

"Sodium Hypochlorite." Medline Plus. http://www.nlm.nih.gov/medlineplus/ency/article/002488.htm (accessed on November 8, 2005).

$$Na^+ \quad O^- \diagdown O \diagdown B \diagdown O$$

Sodium Perborate

OVERVIEW

Sodium perborate (SO-dee-um per-BOR-ate) is a white amorphous powder commonly available as the monohydrate ($NaBO_3 \cdot H_2O$) or the tetrahydrate ($NaBO_3 \cdot 4H_2O$). The most frequently available form of the compound, the tetraborate, is a white crystalline solid with a salty taste that melts at 63°C (145°F) and loses its water of hydration when heated above 130°C (270°F). When dissolved in water, all forms of sodium tetraborate decompose to yield hydrogen peroxide (H_2O_2) and sodium borate ($Na_2B_4O_7$). The formation of hydrogen peroxide, which is itself unstable and breaks down to release nascent oxygen (O), makes sodium tetraborate an excellent source of "active" oxygen. The terms *nascent* and *active* refer to individual atoms of oxygen that have a strong tendency to react with other elements and compounds. Since sodium tetraborate is more stable than hydrogen peroxide, it can be used for the same purposes as the peroxide, but is safer and easier to handle. The most common applications of sodium tetraborate are in detergents, bleaches, hair care products, and disinfectants.

HOW IT IS MADE

Anhydrous sodium perborate can be made by heating the tetrahydrate:

$$NaBO_3 \cdot 4H_2O \rightarrow NaBO_3 + 4H_2O$$

or by reacting hydrogen peroxide with sodium metaborate:

$$H_2O_2 + NaBO_2 \rightarrow NaBO_3 + H_2O$$

Sodium perborate tetrahydrate is prepared by reacting hydrogen peroxide with borax ($Na_2B_4O_7$).

COMMON USES AND POTENTIAL HAZARDS

Sodium perborates' uses are based on the fact that it is a mild and relatively safe oxidizing agent. An oxidizing agent is a substance that supplies oxygen to other substances. The oxidation of a material may cause bleaching or the destruction of disease-causing microogranisms. For example, sodium perborate is added to some detergents to improve their bleaching capability. It makes the detergents more effective in removing stains, keeping white fabrics white, and preserving the original colors of colored cloth. The compound is

Interesting Facts

- Because of concerns about the hazard that boron compounds may pose to the environment, sodium perborate is being replaced in many applications by a similar but safer oxidizing agent, sodium percarbonate.

also added to some automatic dishwasher powders to improve the product's ability to loosen left-on food and to sterilize dishes, silverware, and cookware. Sodium perborate is also used as an ingredient in a variety of home- and personal-care products, such as deodorants, mouthwashes, denture cleaners, and toothpastes.

Sodium perborate is an irritant to the skin, eyes, and respiratory tract. If ingested, it may cause nausea, vomiting, diarrhea, abdominal pain, and bleeding. Exposure to large quantities of the pure compound can produce severe skin rashes, permanent eye damage, breathing difficulties, unconsciousness, and kidney failure. Despite these concerns, people who use personal- and home-care products containing sodium perborate are at low risk for health problems. The compound degrades during machine washing, and home users rarely ingest, inhale, or come into contact with significant quantities of the compound. The European Union Scientific Committee on Toxicity, Ecotoxicity, and the Environment issued a report in May 2004 stating that manufacturers and consumers do not need to take steps beyond those already in place to protect themselves from exposure to sodium perborate.

FOR FURTHER INFORMATION

"Material Safety Data Sheet." Solvay Interox.
    http://www.solvayinterox.com.au/solvay/uploadfile/
    SOL030%20-%20PBST.doc (accessed on November 10, 2005).

"Opinion on the Results of the Risk Assessment of Sodium Perborate." Scientific Committee on Toxicity, Ecotoxicity, and the Environment (CSTEE), 28 May 2004.
    http://europa.eu.int/comm/health/ph_risk/committees/sct/
    documents/out225_en.pdf (accessed on November 10, 2005).

Words to Know

AMORPHOUS Without crystalline structure.

"Sodium Perborate Information." PAN Pesticides Database. http://www.pesticideinfo.org/Detail_Poisoning.jsp?Rec_Id=PC34416 (accessed on November 10, 2005).

"Sodium Perborate Monohydrate." Shangyuchem. http://www.chem-world.com/sodium-perborate.htm (accessed on November 10, 2005).

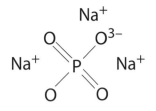

OTHER NAMES:
See Overview

FORMULA:
Monobasic: NaH$_2$PO$_4$;
Dibasic: Na$_2$HPO$_4$;
Tribasic: Na$_3$PO$_4$

ELEMENTS:
Sodium, phosphorus,
oxygen

COMPOUND TYPE:
Salt (inorganic)

STATE:
Solid

MOLECULAR WEIGHT:
119.98 to 163.94 g/mol

MELTING POINT:
Not applicable

BOILING POINT:
Not applicable

SOLUBILITY:
Soluble in water;
insoluble in ethyl
alcohol

K E Y   F A C T S

Sodium Phosphate

OVERVIEW

The three forms of sodium phosphate are formed when one or more of the three hydrogen atoms in phosphoric acid (H$_3$PO$_4$) are replaced by sodium atoms. When one hydrogen is replaced, the monobasic form is produced; replacing two hydrogen atoms results in the formation of the dibasic form; and replacing all three hydrogens results in the formation of tribasic sodium phosphate. All three forms of sodium phosphate are colorless to white crystalline solids or white powders. All may occur as hydrates, such as monobasic sodium phosphate monohydrate and dihydrate (NaH$_2$PO$_4$·H$_2$O and NaH$_2$PO$_4$·2H$_2$O); dibasic sodium phosphate dihydrate, hepta-hydrate, and dodecahydrate (Na$_2$HPO$_4$·2H$_2$O, Na$_2$HPO$_4$·7H$_2$O, and Na$_2$HPO$_4$·12H$_2$O); and tribasic sodium phosphate dodeca-hydrate (Na$_3$PO$_4$·12H$_2$O).

The three salts are also known by a number of other names, such as: NaH$_2$PO$_4$: primary sodium phosphate, primary sodium orthophosphate, sodium biphosphate, MSP; Na$_2$HPO$_4$: second-ary sodium phosphate, secondary sodium orthophosphate,

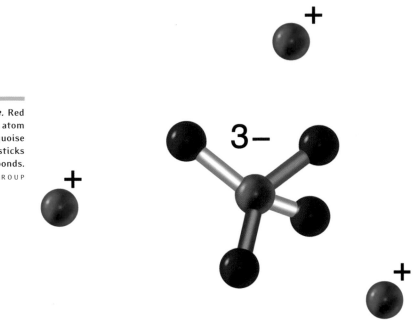

Sodium phosphate. Red atoms are oxygen; orange atom is phosphorus; and turquoise atoms are sodium. Gray sticks indicate double bonds.
PUBLISHERS RESOURCE GROUP

disodium phosphate, disodium hydrogen phosphate, DSP; Na_3PO_4: tertiary sodium phosphate, tertiary sodium orthophosphate, trisodium phosphate, TSP.

HOW IT IS MADE

All forms of sodium phosphate are made by treating phosphoric acid with a sodium compound to replace one or more of the hydrogen atoms in the acid. For example, reacting phosphoric acid with sodium hydroxide will make monobasic sodium phosphate:

$$H_3PO_4 + NaOH \rightarrow NaH_2PO_4 + H_2O$$

Reacting phosphoric acid with sodium carbonate will produce the dibasic form of the compound:

$$H_3PO_4 + Na_2CO_3 \rightarrow Na_2HPO_4 + H_2O + CO_2$$

And treating phosphoric acid with an excess of sodium hydroxide will result in the formation of the tribasic form of the compound:

$$H_3PO_4 + 3NaOH \rightarrow Na_3PO_4 + 3H_2O$$

Interesting Facts

- Chicken processors often dip whole chickens in a solution of TSP before treating and packaging them. The compound kills bacteria and reduces the risk of food poisoning.

Other methods of preparation are available for all three forms of sodium phosphate.

COMMON USES AND POTENTIAL HAZARDS

The three forms of sodium phosphate have somewhat different uses. Monobasic sodium phosphate is used as a food additive to maintain proper acidity and in baking powders, as a food supplement to provide the phosphorus needed in a person's daily diet, in the treatment of boiler water to reduce the formation of scale on the inner surface of the boiler, and as a feed supplement for cattle and other farm animals.

Dibasic sodium phosphate is used as a food additive to maintain emulsions and proper acidity of food products, in the manufacture of fertilizers, as a food supplement for humans and farm animals, in the treatment of silk, for fireproofing of wood and paper products, to treat boiler water, in the production of detergents, as a raw material in the manufacture of ceramics, as a mordant for dyeing, and as a cathartic and laxative.

Tribasic sodium phosphate is used in a variety of cleaning agents, such as detergents, industrial cleaning products, and metal cleaners; to treat boiler water; for the tanning of leather; in the manufacture of textiles and paper products; in the purification of sugar; as a dietary supplement for humans and farm animals; in paint removers; and for various photographic purposes.

All three forms of sodium phosphate are mild skin, eye, and respiratory system irritants. Tribasic sodium phosphate has somewhat more serious health consequences than the monobasic or dibasic forms. Exposure to the compounds may produce redness, itching, burns, pain, and blisters on

Words to Know

CATHARTIC A substance that promotes bowel movements.

EMULSION A temporary mixture of two liquids that normally do not dissolve in each other.

MORDANT A substance used in dyeing and printing that reacts chemically with both a dye and the material being dyed to help hold the dye permanently to the material.

the skin; redness, pain, and burning of the eyes; and a burning sensation, coughing, shortness of breath, and a sore throat if ingested. Taken in larger amounts, the tribasic form may cause vomiting, nausea, and diarrhea and may induce shock and collapse. In such cases, immediate medical assistance is required.

FOR FURTHER INFORMATION

"Sodium Phosphate Dibasic Heptahydrate." Ted Pella, Inc. http://www.tedpella.com/msds_html/19545msd.htm (accessed on November 10, 2005).

"Sodium Phosphate, Monobasic." Cornell University. http://msds.ehs.cornell.edu/msds/msdsd0d/a9/ m4481.htm#Section3 (accessed on November 10, 2005).

"Sodium Phosphates." Agricultural Marketing Service, U.S. Department of Agriculture. http://www.ams.usda.gov/nop/NationalList/TAPReviews/ sodiumphosphates.pdf (accessed on November 10, 2005).

"Trisodium Phosphate." Household Products Database, National Institutes of Health. http://householdproducts.nlm.nih.gov/cgi-bin/household/ brands?tbl=brands&id=3007033 (accessed on November 10, 2005).

"Trisodium Phosphate (Anhydrous)." International Chemical Safety Cards. http://www.inchem.org/documents/icsc/icsc/eics1178.htm (accessed on November 10, 2005).

"USDA Approves Phosphate to Reduce Salmonella in Chicken." *Environmental Nutrition* (February 1993): 3.

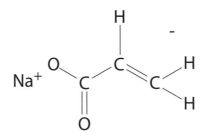

KEY FACTS

Sodium Polyacrylate

OVERVIEW

Sodium polyacrylate (SO-dee-um pol-ee-AK-ruh-late) is an odorless, grainy white powder. Its most impressive property is its ability to absorb large amounts of fluid, up to 800 times its volume of distilled water and lesser amounts of other liquid mixtures. This property accounts for one of its primary applications, in the manufacture of disposable diapers. Diapers made from sodium polyacrylate are able to absorb up to 30 grams of urine for each gram of diaper.

HOW IT IS MADE

Sodium polyacrylate is produced by the reaction between acrylic acid (H$_2$C=CHCOOH) and its sodium salt (H$_2$C=CHCOONa). The product of this reaction is a long-chain copolymer consisting of alternate units of acrylic acid and sodium acrylate. A copolymer is a polymer made of two different monomers, in this case, acrylic acid and sodium acrylate. What makes this polymer different from most other

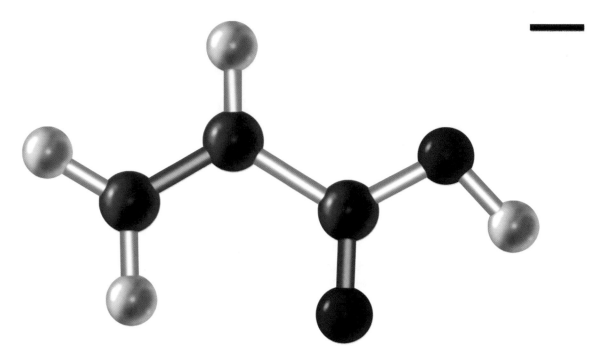

Sodium polyacrylate. Red atoms are oxygen; white atoms are hydrogen; black atoms are carbon; turquoise atom is sodium. Gray sticks indicate double bonds. PUBLISHERS RESOURCE GROUP

kinds of polymers is that adjacent polymer chains are able to cross link with each other. The hydrogen on a carboxyl group (-COOH) on one chain reacts with a double bond (-C=C-) on an adjacent chain, forming a link that holds the two chains together. Cross-linking occurs at many points in the polymer, resulting in the formation of a mesh-like web consisting of polymer chains.

When water is added to the polymer, it forces carboxyl groups away from each other, forcing the mesh to open up

Interesting Facts

- Sodium polyacrylate was first developed by researchers for the National Aeronautics and Space Administration. The material was used for diapers worn by astronauts while they were on long space trips.

and make space for water molecules to fill the gaps in the polymer. As more water is added, the carboxyl groups stretch even farther apart, making room in turn for yet more water molecules to be absorbed by the polymer. If the polymer is allowed to dry out, water molecules leave gaps in the compound, the empty spaces between carboxyl groups collapse, and the polymer returns to its original size. It can then be stored and re-used any number of times.

COMMON USES AND POTENTIAL HAZARDS

The primary use of sodium polyacrylate is in the manufacture of baby diapers. The need for some sort of disposable diaper first arose during World War II because of a shortage of cotton, from which cloth diapers are made. The compound now has a number of other applications. For example, it is used to pack poultry, red meat, fish, fresh-cut fruits and vegetables, and fresh whole berries to keep them moist and fresh. Fluids from washing these foods during processing may accumulate inside a package and provide an environment for the growth of bacteria that cause foods to spoil. A packaging material made of cellulose and sodium polyacrylate absorbs these fluids and prevents them from being squeezed out of the package.

Sodium polyacrylate is also used as a thickening agent in medical gels used to treat bed sores, which are open wounds that develop when a person is bed-ridden for too long. The compound is also added to detergents and to potting soils to help retain water. The compound is now being used in some parts of the world where there is insufficient rain to allow

Words to Know

COPOLYMER A polymer that consists of two or more different monomers.

POLYMER A compound consisting of very large molecules made of one or two small repeated units called monomers.

crops or lawns to grow. It absorbs moisture when rain does fall and holds it in place until plants can absorb the water.

In industry, sodium polyacrylate is used in filtration units that remove water from airplane and automotive fuel. It is also used as a thickening agent in coatings and adhesives used in the upholstery, drapery, carpet, paper, paint, wallpaper, printing, and textile industries. The compound is also used to thicken certain liquid products applied by spraying, such as cleaning products. Finally, it is sometimes used to prevent fluid loss in oil wells.

Sodium polyacrylate is a mild irritant to the skin, eyes, and respiratory tract. It may cause redness, itching, and pain on the skin or in the eyes; and coughing, shortness of breath, and inflammation of the respiratory tract. Questions have been raised about possible health effects on babies who wear disposable diapers containing sodium polyacrylate. Some people suggest that a baby's tender skin may be more sensitive to the irritation caused by sodium polyacrylate than the skin of an adult. The compound was removed from tampons in 1985 because some women who left their tampons in place too long experienced unacceptable levels of irritation caused by sodium polyacrylate in the product.

FOR FURTHER INFORMATION

Allison, Cathy. "Disposable Diapers: Potential Health Hazards?" BCParent Online. http://www.bcparent.com/articles/baby_talk/disposable_diapers.html (accessed on November 10, 2005).

"Environmental Assessment." U.S. Food and Drug Administration, Center for Food Safety and Applied Nutrition. http://www.cfsan.fda.gov/~acrobat2/fnea0427.pdf (accessed on November 10, 2005).

Mebane, Robert C., and Thomas R. Rybolt. *Plastics and Polymers.* New York: Twenty-First Century, 1995.

"Sodium Polyacrylate." Center for Advanced Microstructures and Devices, Louisiana State University. http://www.camd.lsu.edu/msds/s/sodium_polyacrylate.htm (accessed on November 10, 2005).

"Sodium Polyacrylate." Flinn Scientific Company/Department of Chemistry, Iowa State University. http://avogadro.chem.iastate.edu/MSDS/Na_polyacrylate.pdf (accessed on November 10, 2005).

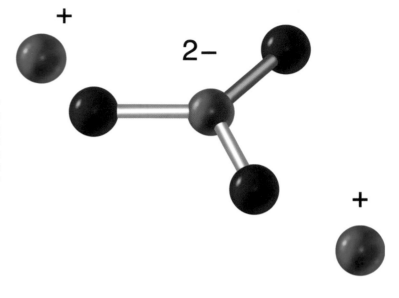

Sodium silicate. Red atoms are oxygen; orange atom is silicon; and turquoise atoms are sodium. Gray stick indicates double bond. PUBLISHERS RESOURCE GROUP

silicates are sometimes referred to as the simplest form of glass.

HOW IT IS MADE

Sodium silicates are made by fusing (melting) sand (silicon dioxide) and soda ash (sodium carbonate) or sodium hydroxide in a gas-fired open hearth furnace, somewhat similar to the furnaces used in the manufacture of steel. The products of this reaction are lumps of sodium silicate that are broken apart and dissolved in a stream of hot steam. The proportions of sand and soda ash used, the temperature of the reaction, and the amount of water that remains in the final product all determine the physical properties of the final product.

COMMON USES AND POTENTIAL HAZARDS

About 1.1 million metric tons (1.3 million short tons) of sodium silicate was produced in the United States in 2004. The primary application for the compound is in the manufacture of soaps and detergents. It improves the cleaning ability of these products and is less damaging to metal components of dishwashers and washing machines than other ingredients of soaps and detergents. Sodium silicate

Interesting Facts

- One of the early uses of sodium silicate was as a preservative for eggs. Eggs were soaked in solutions of sodium silicate, or the compound was painted on the egg shells. The sodium silicate filled the pore in the egg shell, preventing bacteria from entering the egg and causing it to spoil. The process was very effective and preserved eggs for up to nine months. More efficient and less expensive methods of preserving eggs are now available.

is also used as a water softener, used by itself or as an ingredient in detergents.

The next most important applications of sodium silicate are as catalysts and in the pulp and paper industry. Catalysts are materials that increase the rate of a chemical reaction without undergoing any change in their own chemical structure. In the pulp and paper industry, sodium silicate is used to bleach raw pulp and help remove ink from scrap paper being reprocessed. The compound is also used in sealants and adhesives. Some other applications of sodium silicate include:

- For the purification of water in municipal and industrial water treatment plants;

- For the fireproofing of fabrics;

- As an anti-caking agent in food products;

- As an additive in cements, where it helps the cement set more quickly;

- In the manufacture of other compounds of silicon by the chemical industry;

- In fluids used to lubricate drilling instruments;

- As a liner for chemical and industrial equipment, such as furnaces used to make steel;

Words to Know

MUCOUS MEMBRANES The soft tissues that line the breathing and digestive passages.

VISCOUS Describing a syrupy liquid that flows slowly.

- In the processing of ores from which metals are extracted; and

- As a binder on grindstones and abrasive wheels.

Sodium silicates are strong irritants to the skin, eyes, and respiratory system. Prolonged exposure to sodium silicate dust, powder, or liquid may cause inflammation of the skin, eyes, nose, and throat. More serious symptoms may include difficulty in swallowing, burns inside the stomach, damage to the mucous membranes, rapid heartbeat, hypertension, shock, severe damage to the lining of the gastrointestinal tract, various types of cancer, and death. These hazards are of concern primarily to workers who come into contact with sodium silicate in solid or liquid form in the workplace.

FOR FURTHER INFORMATION

Higgins, Kevin T. "Simplified Food-Oil Refining." *Food Engineering* (February 1, 2003). Also available online at http://www.foodengineeringmag.com/CDA/ArticleInformation/features/BNP__Features__Item/0,6330,94941,00.html (accessed on November 10, 2005).

"Practical Uses for Sodium Silicate." The Chemistry Store.com. http://www.chemistrystore.com/sodium_silicate_uses.htm (accessed on November 10, 2005).

"Silicate Chemistry." PQ Corporation. http://www.pqcorp.com/technicalservice/understanding_silicatesolchem.asp (accessed on November 10, 2005).

"Sodium Metasilicate." International Programme on Chemical Safety. http://www.inchem.org/documents/pims/chemical/

pim500.htm#SectionTitle:1.3%20%20Synonyms (accessed on November 10, 2005).

"Sodium Silicates." Chemical Land 21. http://www.chemicalland21.com/arokorhi/industrialchem/ inorganic/SODIUM%20SILICATE.htm (accessed on November 10, 2005).

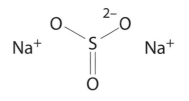

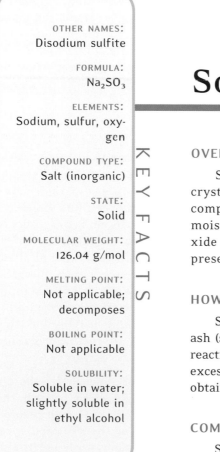
Sodium Sulfite

OVERVIEW

Sodium sulfite (SO-dee-um SUL-fite) is a white powder or crystalline solid with no odor but a slightly salty taste. The compound is stable in dry air, but tends to decompose in moist air to produce sulfur dioxide (SO$_2$) and sodium hydroxide (NaOH). The compound has a variety of uses as a food preservative and in the paper and pulp industry.

HOW IT IS MADE

Sodium sulfite can be prepared by reacting sulfur dioxide, soda ash (sodium carbonate; Na$_2$CO$_3$), and water. The product of this reaction is sodium bisulfite (NaHSO$_3$), which is then treated with excess soda ash to obtain sodium sulfite. The compound can also be obtained as a byproduct in the preparation of phenol (C$_6$H$_5$OH).

COMMON USES AND POTENTIAL HAZARDS

Sodium sulfite is an essential chemical in the pulp and paper industry. Just over half of all the sodium sulfite made

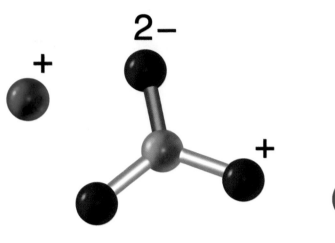

Sodium sulfite. Red atoms are oxygen; yellow atom is sulfur; and turquoise atoms are sodium. PUBLISHERS RESOURCE GROUP

in the United States is used by the pulp and paper industry. The compound acts as a pulping agent for wood, rags, and straw. A pulping agent is a substance that breaks down raw materials and converts them into the pulp from which paper is made. Sodium sulfite is also used to remove excess chlorine used to bleach wood pulp and other raw materials needed in the production of paper.

The second largest application of sodium sulfite is in water and wastewater treatment plants, where it is used to react with and neutralize excess chlorine used in the water and wastewater treatment processes. The third most important application of sodium sulfite is in photography. The compound is used in the developing process, and it acts as a preservative for the final picture produced. Sodium sulfite is still used as a food preservative also, although the conditions under which it can be added are somewhat limited. In addition to its hazards among individual with allergies, sodium sulfite destroys both vitamins B_1 and E, meaning that it cannot be added to foods that contain these vitamins. It is still widely used, however, in the wine-making industry for the control of bacteria involved in the wine-making process.

As noted above, a significant number of people are allergic to sodium sulfite. In addition to this health hazard, the compound can be an irritant to the skin, eyes, and respiratory tract. It can cause inflammation of the skin and eyes, irritation of the nose and throat, problems with breathing, and stomach upset. With the level at which most people come

Interesting Facts

- Sodium sulfite has been used as a food preservative for many years. However, in 1986 the U.S. Food and Drug Administration (FDA) banned the use of sodium sulfites for certain types of food. The agency had discovered that about one in a hundred people are sensitive to sodium sulfite. Thirteen deaths and more than 500 allergic reactions resulting from exposure to sodium sulfite had been reported to the FDA. The agency now prohibits the use of sodium sulfite as a preservative on raw fruits and vegetables. Processed foods that contain sodium sulfite must include a notice to that effect on the food label.

into contact with the compound, however, it poses little threat to a person's health.

FOR FURTHER INFORMATION

"Sodium Sulfite." Esseco General Chemical. http://www.genchemcorp.com/pdf/msds/Sodium%20Sulfite,%20EGC%20-%204-03.pdf (accessed on November 12, 2005).

"Sodium Sulfite." J. T. Baker. http://www.jtbaker.com/msds/englishhtml/s5066.htm (accessed on November 12, 2005).

"Sodium Sulfite." Solvay Chemicals. http://www.solvaychemicals.us/pdf/Sodium_Sulfite/SODSULF.pdf (accessed on November 12, 2005).

"Sodium Sulfite Photographic Grade." Center for Advanced Microstructures and Devices, Louisiana State University. http://www.camd.lsu.edu/msds/s/sodium_sulfite.htm (accessed on November 12, 2005).

See Also Sulfur Dioxide

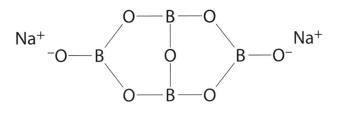

OTHER NAMES:
Sodium borate;
sodium pyroborate;
disodium tetraborate;
borax

FORMULA:
$Na_2B_4O_7$ or
$Na_2B_4O_7 \cdot 10H_2O$; see
Overview

ELEMENTS:
Sodium, boron,
oxygen

COMPOUND TYPE:
Salt (inorganic)

STATE:
Solid

MOLECULAR WEIGHT:
$Na_2B_4O_7$: 201.22 g/
mol; $Na_2B_4O_7 \cdot 10H_2O$:
381.37 g/mol

MELTING POINT:
$Na_2B_4O_7$: 743°C
(1370°F);
$Na_2B_4O_7 \cdot 10H_2O$:
decomposes at about
75°C (170°F)

BOILING POINT:
$Na_2B_4O_7$: 1575°C
(2867°F);
$Na_2B_4O_7 \cdot 10H_2O$: not
applicable

SOLUBILITY:
Soluble in water

KEY FACTS

Sodium Tetraborate

OVERVIEW

Sodium tetraborate (SO-dee-um tet-ruh-BOR-ate) is a term used for either the anhydrous or hydrated form of the compound with the formula $Na_2B_4O_7$. The decahydrate ($Na_2B_4O_7 \cdot 10H_2O$) is also referred to as borax. Borax also occurs without water of hydration and in that form is known as anhydrous borax.

Some historians think that borax may have been known as long as 4,000 years ago. Since ancient people did not use the term, however, there is considerable doubt as to the authenticity of these claims. The compound was certainly in use as far back as about 800 BCE when the compound was being used in the Chinese and Islamic civilization for making glass and in jewelry work. The substance was very expensive, however, and it was not widely used in Europe until the Middle Ages. Borax became more commonly used after extensive deposits of its naturally occurring form were found in the United States. The first of those deposits was discovered in Nevada in 1879, although the largest deposits were later found in the desert regions of southern California. Today, the

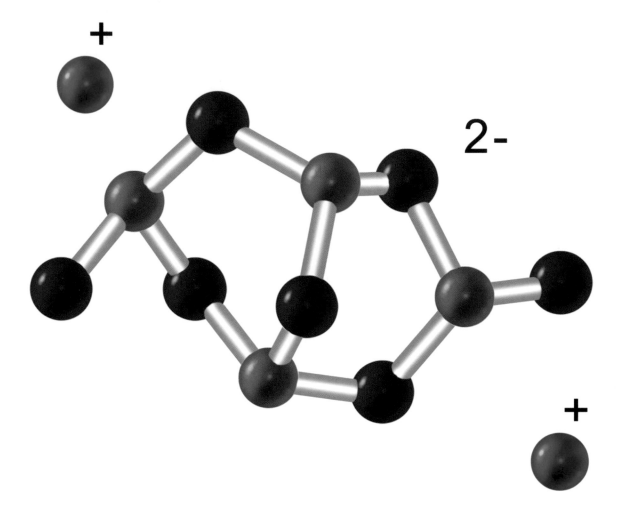

largest reserves of borax are found near the town of Boron, California, and at Borax Lake, California. The compound is also mined in Tibet, Russia, Chile, and Turkey.

Sodium tetraborate is an odorless white crystalline solid or powder. The hydrated form loses its water of hydration when heated and then fuses (melts) to form a glass-like solid at higher temperatures.

HOW IT IS MADE

Sodium tetraborate occurs naturally as the minerals tincal (pronounced "tinkle;" $Na_2B_4O_7 \cdot 10H_2O$) and kernite

Interesting Facts

- Perhaps the best known commercial form of sodium tetraborate is called Twenty-Mule-Team® Borax. The name comes from the fact that the first borax mines in California were located 165 miles from the nearest train station in Mojave, California. The mined borax was transported that distance in wagons pulled by 20 mules. Each wagon cost $900 to build, weighed 14,500 kilograms (32,000 pounds), and had wheels 2 meters (7 feet) in diameter. The trip took about twenty days, often in temperatures as high as 45°C (113°F). Over the six years during which mule teams were used, about nine million kilograms (20 million pounds) of borax were moved from mine to railway station. In 1896 the Pacific Coast Borax Company (later, U.S. Borax) took the twenty-mule teams as their corporate symbol.

($Na_2B_4O_7 \cdot 4H_2O$). Ores containing these minerals are crushed, washed, and processed to obtain the decahydrate of high purity. Anhydrous sodium tetraborate can be obtained by heating the decahydrate. Sodium tetraborate can also be obtained by processing other minerals that contain borates, such as ulexite ($NaCaB_5O_9 \cdot 8H_2O$) and colemanite ($Ca_2B_6O_{11} \cdot 5H_2O$).

COMMON USES AND POTENTIAL HAZARDS

The primary use of sodium tetraborate is in the manufacture of glass products. About 43 percent of all the compound used in the world goes to this application. Glass made with sodium tetraborate is very strong and heat resistant. The well-known Pyrex® brand of glass is made with sodium tetraborate. Today, the largest single use of borax glass is in the manufacture of fiberglass insulation and fiberglass textiles.

The next most important use of sodium tetraborate is in the manufacture of soaps, detergents, and personal care products. Some well known products that contain borax

Words to Know

ANHYDROUS Describing a compound that lacks any water of hydration.

DECAHYDRATE Form of a compound that exists with ten molecules of water.

FLUX A material that aids the processes of welding and soldering (joining) metals.

WATER OF HYDRATION Water that has combined with a compound by some physical means.

include 20-Mule-Team® Borax all-purpose cleaner, 20-Mule-Team® Borax Laundry Booster, Borateem® stain remover, and Boraxo® powdered hand soap. Some other uses of sodium tetraborate include:

- As a flame-retardant and fungicide for wood products;
- In the production of enamel, porcelain, glazes, enamels, and frits (specialized types of glass);
- In the manufacture of fertilizers and herbicides;
- As additives for certain kinds of polymers;
- As a flux for smelting and soldering metals;
- In the preparation of rust inhibitors; and
- In certain photographic processes.

Sodium tetraborate is a mild irritant to the skin, eyes, and respiratory tract. It can cause inflammation of the skin, eyes, nose, throat, and lungs. If ingested, it can cause nausea, vomiting, diarrhea, and abdominal pain. There is some evidence that ingestion of sodium tetraborate can cause reproductive problems in laboratory animals, although similar effects have not been seen in humans. Swallowing large quantities of sodium tetraborate can have serious health consequences, especially for young children. The fatal dose of sodium tetraborate for young children is about five grams (0.2 ounce) of the compound.

FOR FURTHER INFORMATION

"Borax (Hydrated Sodium Borate)." Amethyst Galleries, Inc. http://mineral.galleries.com/minerals/carbonat/borax/borax.htm (accessed on November 12, 2005).

"Rio Tinto Borax." Borax.
http://www.borax.com/index.html (accessed on November 12, 2005).

"Sodium Tetraborate." International Programme for Chemical Safety.
http://www.inchem.org/documents/icsc/icsc/eics1229.htm (accessed on November 12, 2005).

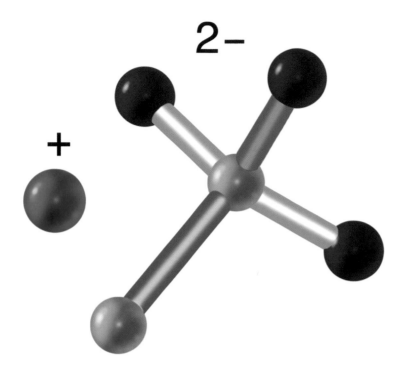

Sodium thiosulfate. Red atoms are oxygen; yellow atoms are sulfur; and turquoise atoms are sodium. Gray sticks indicate double bonds.
PUBLISHERS RESOURCE GROUP

of sodium sulfide (NaS$_2$) and dyes that contain sulfur. The compound can also be produced directly by adding powdered sulfur to a solution of sodium sulfite and heating the reactants:

$$Na_2SO_3 + S \rightarrow Na_2S_2O_3$$

COMMON USES AND POTENTIAL HAZARDS

In addition to its use as an antichlor, sodium thiosulfate can, itself, also be used as a bleach for paper, pulp, bone, straw, ivory, and other materials. Its other major application is in photography, where it is used as a fixing agent. A fixing agent is a chemical that reacts with silver bromide and silver chloride on a photographic film that has not been exposed. The silver bromide and silver chloride are then washed away, leaving behind and "fixed in place" the silver that has been produced by exposure to light. Smaller amounts of sodium thiosulfate are used for other purposes, such as:

• A food additive for the purpose of maintaining the proper of acidity of a food product or sequestering

Interesting Facts

- An important, but seldom used, application of sodium thiosulfate is as an antidote for cyanide poisoning.

(capturing and holding) unwanted materials in the food;

- For certain medical purposes, such as the treatment of fungal infections of the skin in humans and, specifically, for the treatment of ringworm in animals;

- In hide tanning and dyeing procedures that use compounds of the element chromium;

- For the extraction of silver metal from its ores; and

- In the analysis of the composition of chemical mixtures.

Sodium thiosulfate is an irritant to the skin, eyes, and respiratory tract, although such effects tend to be mild unless one is exposed to large quantities of sodium thiosulfate dust, mist, or solutions. The compound can also be toxic if ingested, causing irritation of the gastrointestinal tract. The amount of sodium thiosulfate permitted in foods is 0.1 percent. No serious long-term effects of exposure to the compound are known.

Words to Know

ANTICHLOR A chemical that reacts with excess chlorine used for purification, disinfecting, or some other purpose.

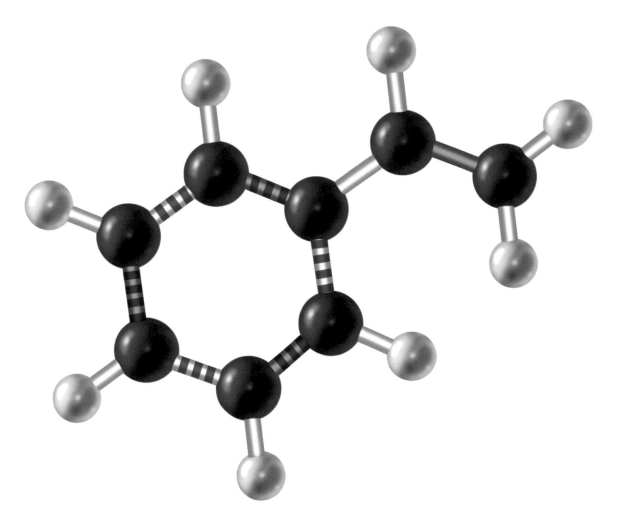

over a catalyst of iron(III) oxides at temperatures of about 600°C (1100°F). Dehydrogenation is the process by which hydrogen atoms are removed from a compound:

$$C_6H_5CH_2CH_3 \cdot H_2 \rightarrow C_6H_5CH=CH_2$$

COMMON USES AND POTENTIAL HAZARDS

About two-thirds of all the styrene produced in the United States is used in the manufacture of polystyrene. Polystyrene is a clear, colorless, hard plastic that is easily molded and made into a foam known as styrofoam. It is used in the insulation of electrical wires and devices, in containers for hot and cold foods and drinks, and for the insulation of buildings.

Interesting Facts

- In 2004, about 5.4 million metric tons (5.9 million short tons) of styrene were produced in the United States. It ranked 17th among all chemicals produced that year.

Almost all of the remaining styrene is used in the production of other polymers, such as acrylonitrile-butadiene-styrene resins, styrene-acrylonitrile resins, styrene-butadiene rubber and latex, and various polyester resins.

The styrene fumes to which a person might be exposed are mild irritants to the skin, eyes, and respiratory tract. They may cause inflammation of the skin, eyes, nose, and throat. In larger amounts, they may adversely affect the nervous system causing nausea, tiredness, muscle weakness, depression, and concentration problems. Adequate studies on the carcinogenic properties of styrene have not been conducted, but the International Agency for Research on Cancer has called styrene a possible carcinogen. There are also no studies on the possible reproductive effects of exposure to styrene. In any case, the individuals most at risk for the health hazards of styrene are people who come into contact with the chemical in the workplace.

Styrene is also a moderate fire risk. In certain concentrations, it is also explosive. Again, these risks are of concern primarily to individuals who work with the pure compound in their jobs.

Words to Know

CARCINOGEN A substance that causes cancer in humans or other animals.

RESIN A natural or artificial soft solid material that is used as glue, ink, or in various plastics.

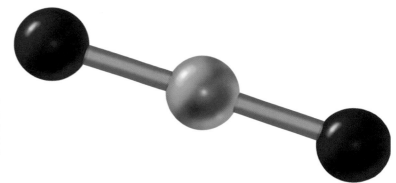

Sulfur dioxide. Red atoms are oxygen and yellow atom is sulfur. PUBLISHERS RESOURCE GROUP

HOW IT IS MADE

Sulfur dioxide can be prepared by several methods, the most common of which is the combustion of sulfur or pyrites (FeS_2). A variety of furnaces have been developed for carrying out this reaction. Each type of furnace produces sulfur dioxide of different purities. After production, the sulfur dioxide is normally cooled and compressed to convert it to liquid form. Liquid sulfur dioxide is more easily stored and transported than the gaseous form. Sulfur dioxide is also obtained as the byproduct of a number of industrial operations, especially the smelting of metallic ores. Smelting is the process by which a metal is extracted from its ore by heating in air. Since many ores are sulfides, this process often results in the formation of sulfur dioxide, which can be captured as a byproduct of the operation. Finally, sulfur dioxide can be produced by the direct combustion of sulfur itself:

$$S + O_2 \rightarrow SO_2$$

COMMON USES AND POTENTIAL HAZARDS

About 45 percent of all the sulfur dioxide produced in the United States is used in the manufacture of other chemical compounds, the most important of which is sodium bisulfite ($NaHSO_3$). Other compounds made from sulfur dioxide include sulfuric acid (H_2SO_4), chlorine dioxide (ClO_2), sodium dithionate ($Na_2S_2O_6 \cdot 2H_2O$), and sodium thiosulfate ($Na_2S_2O_3 \cdot 5H_2O$). Sulfur dioxide is also used as a bleaching agent for a number of products, including pulp and paper, textile fibers, straw, glue, gelatin, starches, grains, and various oils. The compound

Interesting Facts

- Sulfur dioxide was first studied in detail by the English physicist and chemist Joseph Priestley (1733-1804), who invented a method for collecting gases over water.

- The ancient Greeks and Romans fumigated their homes by burning sulfur. The sulfur dioxide formed destroyed microorganisms that cause disease and rot.

- The concentration of sulfur dioxide in clean air above the continents is less than one part per billion. Volcanic eruptions account for about half of all the gas produced by natural sources.

has a number of agricultural uses, especially in the treatment of soybeans and corn to destroy molds and preserve the product from decay. It is also used in the processing and refining of metal ores and petroleum. For example, it is added to some petroleum products to remove dissolved oxygen that would cause rust of pipes through which the products are distributed. Some other applications of the compound include:

- As a preservative for certain dried fruits and vegetables, such as cherries and apricots;

- As an additive in beers to prevent the formation of harmful products known as nitrosamines;

- As an additive in wines to prevent the growth of undesirable molds and other fungi;

- In the production of high-fructose corn syrups (HFCS) used as sweeteners in commercial food and drink products;

- As an antichlor in water purification systems;

- In the refining of sugar;

- In the manufacture of certain clay products to counteract the presence of compounds or iron and other metals that would impart color to the final product;

Words to Know

AMORPHOUS Without crystalline structure.

ELECTROPLATING A process by which a thin layer of one metal is deposited on top of a second metal by passing an electric current through a solution of the first metal.

MORDANT A substance used in dyeing and printing that reacts chemically with both a dye and the material being dyed to help hold the dye permanently to the material.

MUTAGEN A substance that causes a mutation in plants or animals. Mutations are changes in an organism's genetic composition.

PRECIPITATE A solid material that settles out of a solution, often as the result of a chemical reaction.

the tanning of animal hides, the manufacture of specialized inks, and the treatment of minor cuts and bruises. Its medical applications are based on its astringent properties. An astringent is a compound that triggers a loss of water from tissue, thereby causing the tissues to shrink and contract. The compound also has a number of other commercial and industrial uses, including:

- As a mordant in the coloring of fabrics and the printing of colored papers;

- In the manufacture of many kinds of chemicals, including tannates (compounds of tannic acid and a metal), gallic acid ($C_6H_2(OH)_3COOH$; with many of the same uses as tannic acid), and pyrogallic acid ($C_6H_3(OH)_3$; also with similar uses);

- As a clarifying (purifying) agent in the manufacture of beers and wines;

- In electroplating;

- In the manufacture of artificial horn and tortoise shell products;

- As a coagulant in the manufacture of rubber to precipitate out impurities;

- In the deodorizing of crude oil; and

- In photographic processes.

Tannic acid is a mild irritant of the skin, eyes, and respiratory tract. It is toxic by ingestion. In large doses, it may cause liver damage. No evidence is available on its carcinogenic or mutagenic effects in humans.

FOR FURTHER INFORMATION

Eusman, Elmer. "The Ink Corrosion Website." http://www.knaw.nl/ecpa/ink/ (accessed on November 15, 2005).

"Material Safety Data Sheet: Tannic Acid." GFS Chemicals. https://gfschemicals.com/Search/MSDS/2415MSDS.PDF (accessed on November 15, 2005).

Meyer, John R. "By Gall-y." http://www.cals.ncsu.edu/course/ent591k/gally.html (accessed on November 15, 2005).

"Tannic Acid/Tannins." J. T. Baker. http://www.jtbaker.com/msds/englishhtml/t0065.htm (accessed on November 15, 2005).

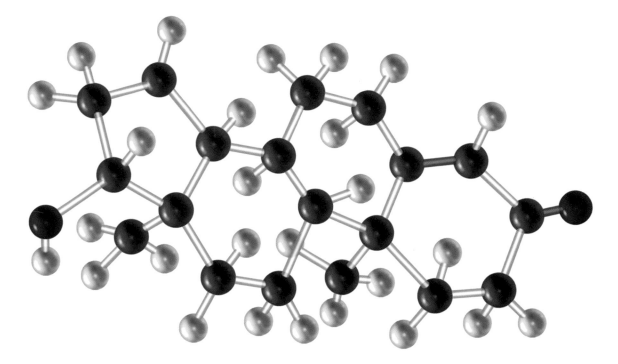

Testosterone. Red atoms are oxygen; white atoms are hydrogen; and black atoms are carbon. Gray sticks indicate double bonds. PUBLISHERS RESOURCE GROUP

laughed off his discoveries, and studies of testosterone were largely suspended for almost half a century.

In the 1930s, however, interest in the compound was revived. In 1935, scientists obtained the first pure sample from testosterone and were able to confirm Brown-Séquard's findings. Within a matter of years, the compound became especially popular among middle-aged men who believed that it could restore their physical and mental abilities. Testosterone was first synthesized in 1935 by German chemist Adolf Friedrich Johann Butenandt (1903–1995), an accomplishment for which he was awarded the 1939 Nobel Prize in Chemistry.

Testosterone is also known by the following names; 17β-Hydroxyandrost-4-ene-3-one; androst-4-en-17β-ol-3-one; testosteroid; testosteron; testostosterone; and trans-testosterone.

HOW IT IS MADE

Testosterone is produced naturally in the male testes and the female ovaries. It is also made synthetically starting either with cholesterol or diosgenin, a plant steroid.

Interesting Facts

- Women who choose to undergo a series of procedures to change their gender are required to take testosterone over a period of many months to stimulate the development of male sex characteristics. They must continue taking testosterone supplements for the rest of their lives.

COMMON USES AND POTENTIAL HAZARDS

Both males and females of all vertebrate species produce testosterone. The amount present in the male body is considerably greater than that present in the female body. Testosterone has a number of biological effects on the body, including an increase in the number of red blood cells and muscles cells and initiation of the development of male sex organs. It is also responsible for the development of secondary male sexual characteristics, such as the growth of body and facial hair, deepening of the voice, and increased sexual desire. Some less desirable effects are an increase in oiliness of the skin and acne.

Testosterone production in men tends to increase during childhood and reaches a maximum during the late teens or early twenties. It then decreases throughout the rest of a man's life. A sudden or extreme decrease in testosterone levels, caused by disease or injury to the hypothalamus, pituitary gland, or testes, can lead to a medical condition known as hypogonadism.

Treatments for hypogonadism in the form of testosterone injections, tablets, skin patches, and skin gels are available. Some men use these devices in an attempt to revive masculine traits that begin to decline normally as one grows older. Many individuals believe, like Brown-Séquard, that testosterone can be something of a "miracle drug" that will restore their lost youth. It has at times been recommended also as a treatment for a host of medical problems, including infertility, impotence, lack of sex drive, osteoporosis,

Words to Know

ANDROGENIC STEROID A hormone responsible for the development of male sexual characteristics.

DERIVATIVE A chemical whose structure is based on or related to another chemical.

HORMONE A chemical produced by living cells in a body that promotes the activity of other cells in the body.

SYNTHESIS A chemical reaction in which some desired chemical product is made from simple beginning chemicals, or reactants.

shortness of stature, anemia, and low appetite. Testosterone may or may not be helpful in treating any one of these conditions.

Since the 1950s, athletes have been using testosterone, its derivatives, and related compounds to improve their performance. The compound increases a person's bone and muscle mass, significantly improving his or her strength and endurance. Testosterone was first used on a large scale basis by athletes from the Soviet Union in the 1950s as part of that nation's efforts to become dominant in world sports.

One problem with using testosterone as a performance-enhancing drug is its undesirable side effects. It tends to increase the size of a man's prostate gland and decrease the size of his testicles. It may also produce wide mood swings that may include dangerously aggressive feelings.

Because of these side effects, researchers have developed compounds that produce the same effects as those obtained from natural testosterone, but without the compound's harmful side effects. Some of those compounds increase the natural production of testosterone in the body when they are ingested. Others are derivatives of testosterone, compounds with similar chemical structures, but minor changes that reduce side effects. These derivatives include compounds such as dihydrotestosterone, androstenedione (also known as andro), dehydroepiandrosterone (DHEA), clostebol, and nandrolone. Today most derivatives of testosterone and testosterone-producing compounds are banned by sports organizations, both because of their harmful side effects and because of the unfair advantages they provide athletes who use them.

FOR FURTHER INFORMATION

Hellstrom, Wayne J. G. "Testosterone Replacement Therapy." *Digital Urology Journal.* http://www.duj.com/Article/Hellstrom2/Hellstrom2.html (accessed on November 15, 2005).

"Material Safety Data Sheet: Testosterone." Paddock Laboratories, Inc. http://www.paddocklabs.com/forms/msds/testost.pdf (accessed on November 15, 2005).

"Testosterone." International Programme on Chemical Safety. http://www.inchem.org/documents/pims/pharm/pim519.htm (accessed on November 15, 2005).

"Testosterone Deficiency." Urology Channel. http://www.urologychannel.com/testosteronedeficiency/index.shtml (accessed on November 15, 2005).

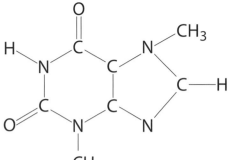

Theobromine

OVERVIEW

Theobromine (thee-oh-BROH-meen) is a white crystalline solid that occurs naturally in cocoa beans, from which chocolate is obtained, and, in smaller amounts, in tea and cola nuts. Theobromine is structurally very similar to caffeine, which differs only in the presence of a methyl group (-CH_3) on one of the nitrogen atoms in the theobromine molecule. Both theobromine and caffeine belong to a family of organic compounds known as the methylxanthines. Theobromine's effects on the human body are similar to those of caffeine, but about ten times weaker. In addition, caffeine is metabolized more quickly, is addictive, and increases alertness and emotional stress. It may also have serious effects on the central nervous system and the kidneys. By contrast, theobromine produces feelings of well-being, is not addictive, has no effect on the central nervous system, and provides only gentle stimulation to the kidneys. Its effects on the body are much longer-lasting than are those of caffeine.

The amount of theobromine in cocoa beans varies widely, ranging from 10 to 40 milligrams of theobromine per gram

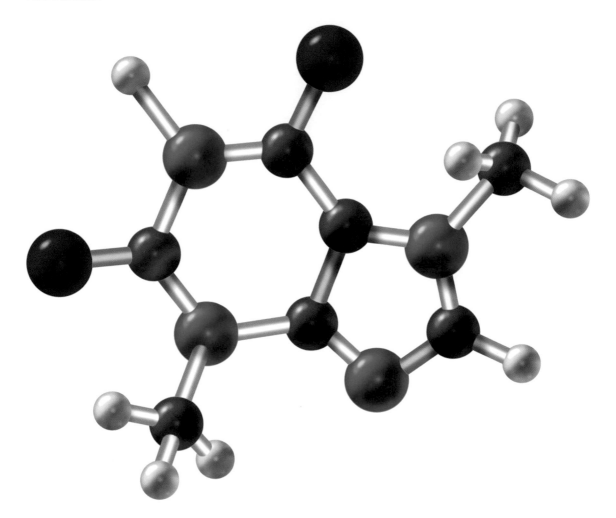

Theobromine. Red atoms are oxygen; white atoms are hydrogen; black atoms are carbon; and blue atoms are nitrogen. Gray sticks indicate double bonds.

of cocoa. The variation depends on a number of facts, including the type of bean, the location where it was grown, and the method of processing the bean. All chocolate products contain theobromine, but the amount varies depending on the type of chocolate. Dark chocolate contains significantly more of the compound than milk chocolate, and high quality chocolate tends to contain more theobromine than low quality chocolate. The characteristic bitter taste of dark chocolate is due to the theobromine present in it.

HOW IT IS MADE

Theobromine is usually obtained from the hulls of cocoa beans left over after the production of chocolate. The hulls

Interesting Facts

- The name *theobromine* is derived from the scientific name for the cacao tree, *Theobroma cacao.*

- Spanish armies that invaded South and Central America in the sixteenth century used chocolate as a source of energy.

- Theobromine is harmless to humans, but very toxic to dogs, horses, and other domestic animals.

are crushed and then treated with an absorbent, such as water or liquid carbon dioxide, which dissolves the theobromine. The water or carbon dioxide is then allowed to evaporate, permitting the crystallization of the pure compound.

COMMON USES AND POTENTIAL HAZARDS

Theobromine occurs in all chocolate products. The pure compound has relatively few uses, however, most of them medical. For example, it has been used as a diuretic—a compound that increases the rate at which liquids are eliminated from the body—and as a mild stimulant. It has also been used to treat hypertension (high blood pressure) because it is a vasodilator, a compound that causes blood vessels to relax

Words to Know

DIURETIC A compound that increases the rate at which liquids are eliminated from the body.

METHYLXANTHINE A compound that is derived from xanthine ($C_5H_4N_4O_2$), with methyl groups (CH_3) replacing one or more of the hydrogen atoms.

STIMULANT: A substance that increases the activity of a living organism or one of its parts.

SUBLIME To go from solid to gaseous form without passing through a liquid phase.

and expand in size. Theobromine also appears to be effective as a cough suppressant, although the quantities needed to achieve useful effects are quite large.

FOR FURTHER INFORMATION

O'Neil, John. "And It Doesn't Taste Bad, Either." *New York Times* (November 30, 2004): F9.

"Theobromine@3Dchem.com." Molecule of the Month. http://www.3dchem.com/molecules.asp?ID=155 (accessed on November 15, 2005).

"What Is Theobromine and What Is Its Effect on Human Beings?" International Cocoa Organization. http://www.icco.org/questions/theobromine.htm (accessed on January 11, 2006).

See Also Caffeine; Carbon Dioxide

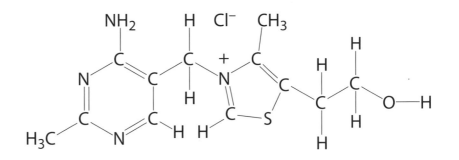

KEY FACTS

Thiamine

OVERVIEW

Thiamine (THYE-uh-min) is the water-soluble vitamin, vitamin B₁. It is also known as 3-(4-Amino-2-methylpyrimidyl-5-methyl)-4-methyl-5,β-hydroxyethylthiazolium. It is usually made available as one of its salts, especially thiamine hydrochloride or thiamine mononitrate. Both salts are white crystalline solids with a bitter taste that are destroyed by alkaline solutions. An alkaline solution is one that contains a strong base.

Thiamine was discovered by Japanese scientist Suzuki Umetaro (1874-1943) in the early twentieth century. Umetaro was investigating a disease known as beriberi that had plagued humans for thousands of years. He found that the disease could be cured by feeding patients a diet that contained rice bran. He was also able to isolate a specific compound in rice bran that produced that effect, a compound he named aberic acid. Aberic acid later became known as thiamine.

The first written reports of beriberi date back 4,000 years. Chinese scholars described a disorder characterized by nausea,

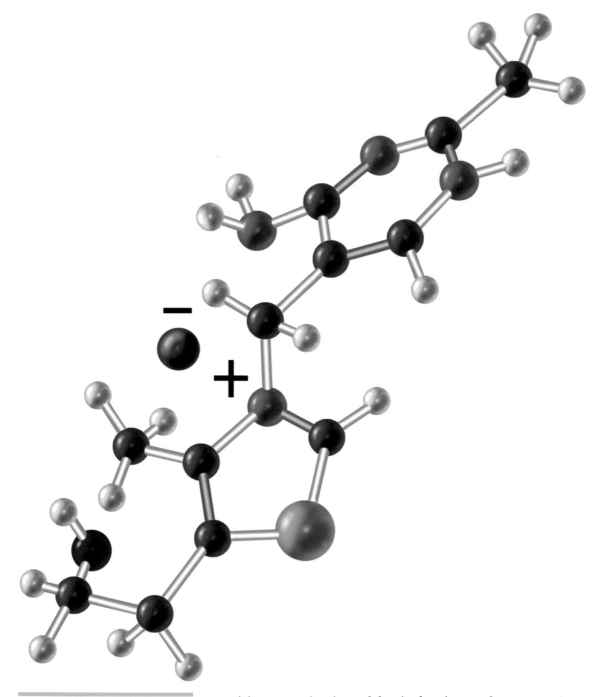

Thiamine. Red atom is oxygen; white atoms are hydrogen; black atoms are carbon; blue atoms are nitrogen; and yellow atom is sulfur. Gray sticks indicate double bonds. PUBLISHERS RESOURCE GROUP

vomiting, constipation, abdominal pain, weakness, wasting, paralysis, irritability, paranoia, depression, and other mental symptoms. In the 1800s, most scientists were convinced that bacteria were responsible for the disease. Over time, it became apparent that this theory was incorrect and that dietary factors

CHEMICAL COMPOUNDS

Interesting Facts

- The word beriberi comes from the Sinhalese expression "I can't, I can't." Sinhalese is the language spoken by natives of Sri Lanka.

- As of 2006, Beriberi is rare in developed countries because of the ready availability of foods and vitamin supplements that contain vitamin B_1. The disease became common in Cuba between 1989 and 1995, however. A trade embargo imposed by the United States reduced the availability of foods and vitamins for the Cuban people.

were responsible for beriberi. In the late nineteenth century, for example, Japanese naval scientist Kanehiro Takaki (1849-1920) noticed that sailors on long voyages developed beriberi if their diets consisted mainly of rice with hulls removed. He also observed that the sailors remained healthy if they included fish, vegetables, wheat, and milk along with rice in their diets.

In 1896, Dutch physician Christiaan Eijkman (1858-1930) found that animals developed beriberi only if they ate polished rice—rice from which hulls had been removed. He learned that the animals recovered if they ate rice with hulls still on it. His colleague Gerrit Grijins (1865-1944) believed that the rice hulls contained some substance that prevented beriberi. By 1910, Umetaro had found that substance, aberic acid.

HOW IT IS MADE

Humans are unable to synthesize thiamine, so they must obtain the vitamin from other sources in which it occurs naturally. The best sources of thiamine are brewer's yeast, whole grains, wheat germ, lean meats, organ meats (such as liver and kidney), fish, dried beans, soybeans, peas, nuts, green leafy vegetables, avocados, raisins, plums, and kelp. Thiamine is also made synthetically by a process that was first developed in the 1930s by American chemist Robert R.

Words to Know

COENZYME A chemical compound that works along with an enzyme to increase the rate at which chemical reactions take place.

SALT An ionic compound where the anion is derived from an acid.

Williams (1886-1965). Williams began his studies of anti-beriberi compounds in 1911 and worked for twenty-five years before discovering the correct chemical formula for the compound. Once he had that information, he was able to invent a system for producing the vitamin artificially. Today, virtually all of the vitamin B_1 sold commercially is prepared artificially.

COMMON USES AND POTENTIAL HAZARDS

Thiamine is a coenzyme needed for a number of essential biochemical reactions in the body. A coenzyme is a chemical compound that works along with an enzyme to increase the rate at which chemical reactions take place. Without enzymes and coenzymes in the body, many chemical reactions would take place so slowly that normal bodily functions could not continue. Thiamine is involved in chemical reactions by which blood is produced and circulated through the body, carbohydrates are metabolized, digestive enzymes are produced, and the nervous system is maintained.

Because it is soluble in water, thiamine is not stored in the body. A person must include the compound in his or her daily diet on a regular basis. Most people ingest adequate amounts of vitamin B_1 in their ordinary diets, and beriberi is very rare in developed countries of the world. It may occur, however, in alcoholics, pregnant women, and people who must undergo kidney dialysis. In all of these cases, a person does not receive adequate amounts of the vitamin for the body's needs. In the case of alcoholics, for example, alcohol replaces the calories they would be getting from food if they were not drinking so much. As a result, they do not get enough vitamin B_1 and other nutrients needed to stay healthy.

The early symptoms of beriberi include fatigue, irritability, poor sleep habits, memory loss, abdominal pain, loss of appetite, and constipation. As the disease progresses, it may produce damage to the peripheral nervous system that serves the arms, legs, feet, and hands, resulting in muscular atrophy (weakness and wasting of the muscles) and loss of sensation in the toes.

FOR FURTHER INFORMATION

Eades, Mary Dan. *The Doctor's Complete Guide to Vitamins and Minerals.* New York: Dell, 2000.

"Thiamine." Medline Plus.
http://www.nlm.nih.gov/medlineplus/ency/article/002401.htm (accessed on November 15, 2005).

"Thiamine and Salts." International Programme on Chemical Safety.
http://www.inchem.org/documents/pims/pharm/pimg015.htm (accessed on November 15, 2005).

"Vitamin B1." Welcome to Glactone.
http://chemistry.gsu.edu/glactone/vitamins/b1/ (accessed on November 15, 2005).

See Also Ascorbic Acid; Cyanocobalamin; Pyridoxine; Riboflavin

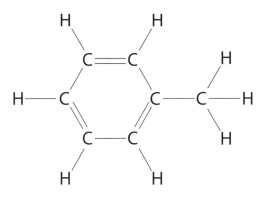

KEY FACTS

Toluene

OVERVIEW

Toluene (TOL-yew-een) is a clear, colorless liquid with a benzene-like odor. It is an aromatic hydrocarbon, that is, a compound that contains carbon and hydrogen only, with the carbon atoms arranged in a ring. Aromatic compounds have structures based on that of benzene (C_6H_6). Toluene was discovered in 1838 by French chemist Pierre Joseph Pelletier (1788-1842). Pelletier found the compound in the gas emitted by the bark of the pine tree *Pinus maritima.* Pelletier named the substance *retinnaphte,* after pine resin. The compound was re-discovered a number of times in later years and given a variety of names, including heptacarbure quadihydrique, benzoene, and dracyl. Toluene's chemical nature was finally determined by German chemist August Wilhelm von Hofmann (1818-1892) and British chemist James Muspratt (1793-1886) who adopted the name *toluol* originally proposed by Swedish chemist Jons Jakob Berzelius (1779-1848).

Words to Know

AROMATIC COMPOUND A compound whose chemical structure is based on that of benzene (C_6H_6).

CATALYTIC REFORMING A process by which hydrocarbons in petroleum are heated, often in contact with a catalyst, in order to change their molecular composition.

MISCIBLE Able to be mixed; especially applies to the mixing of one liquid with another.

SOLVENT A substance that is able to dissolve one or more other substances.

TERATOGENIC Capable of causing an organism to develop incorrectly.

used explosives. Even today, the power of other explosives is measured by comparison with TNT. For example, the power of nuclear weapons is said to be 10 kilotons or 10 megatons, meaning that they have an explosive power comparable to 10,000 tons or 10 million tons, respectively, of TNT.

Toluene is a very hazardous chemical compound. It catches fire easily and, under proper conditions, is explosive. It is also an irritant to the skin, eyes, and respiratory tract. It may produce symptoms such as headache, dizziness, drowsiness, confusion, fatigue, and peculiar sensations of the skin, such as a "pins and needles" effect. It large quantities, toluene may cause unconsciousness, coma, and death. In contact with the skin or eyes, it may produce redness and pain. Prolonged exposure to toluene may cause serious damage to the liver and kidneys resulting in anemia, decreased blood cell count, and bone marrow hypoplasia (diminished ability to produce blood cells). Toluene is thought to be teratogenic, capable of damaging a fetus while still in the womb.

FOR FURTHER INFORMATION

Harte, John, et al. *Toxics A to Z.* Berkeley: University of California Press, 1991, 415-417.

Martin, Kevin A. "Toxicity, Toluene." eMedicine. http://www.emedicine.com/emerg/topic594.htm (accessed on December 29, 2005).

"Toluene." International Chemical Safety Cards.
http://www.cdc.gov/niosh/ipcsneng/neng0078.html (accessed on December 29, 2005).

"Toluene." J. T. Baker.
http://www.jtbaker.com/msds/englishhtml/t3913.htm (accessed on December 29, 2005).

"ToxFAQs™ for Toluene." Agency for Toxic Substances and Disease Registry.
http://www.atsdr.cdc.gov/tfacts56.html (accessed on December 29, 2005).

See Also Benzene

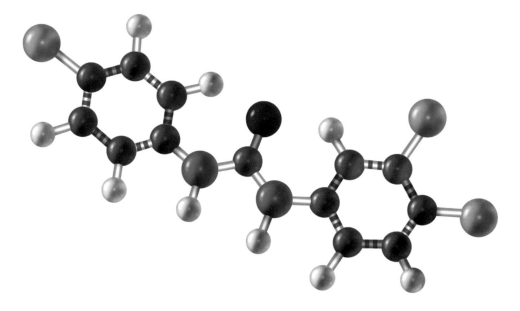

ENR, that many bacteria and fungi use to make cell membranes. If they cannot make cell membranes, these microorganisms die. Humans do not have this enzyme, so triclocarban is considered harmless to them. The use of triclocarban became more widespread after another popular antimicrobial, hexachlorophene, was banned by the U.S. Food and Drug Administration in the 1970s. Hexachlorophene use was discontinued after the antimicrobial was implicated in the death of infants who had been exposed to the product. In the first years of the twenty-first century, triclocarban and triclosan began to be used less, due to increased concerns about the long-term effects of these compounds on the human body and on the environment.

HOW IT IS MADE

Triclocarban is synthesized by one of two procedures. In one, 4-chlorophenyl isocyanate (ClC_6H_4NCO) is reacted with 3,4-dichloroaniline ($C_6H_3NH_2Cl_2$) to give triclocarban. The relationship of these two compounds to the structure of the final product ($C_6H_3Cl_2NHCONHC_6H_4Cl$) is obvious from their chemical structures. In the second method of preparation 3,4-dichlorophenyl isocyanate ($Cl_2C_6H_3NCO$) is reacted with 4-chloroaniline ($ClC_6H_4NH_2$) to give the desired product.

Interesting Facts

- Hunters sometimes wash their clothes in detergents that contain triclocarban to kill the bacteria that contribute to body odor. In this way, the animals on whom they prey, such as deer, are less likely to detect the smell of the hunter and flee the scene.

- No triclocarban is manufactured in the United States. It is imported from other countries, primarily China and India.

COMMON USES AND POTENTIAL HAZARDS

About three-fourths of all liquid soaps and nearly one-third of all bar soap made in the United States contains antimicrobial agents such as triclocarban and triclosan. Research has shown, however, that washing one's hand with plain soap and water provides as much protection from disease-causing organisms as do antimicrobial soaps. It is apparently the action of washing that removes most bacteria. Many public health experts say that antimicrobial soaps are not necessary in normal healthy households.

The growing concern is that the widespread use of antimicrobial soaps may cause harmful bacteria and fungi to become resistant to antimicrobial agents such as triclocarban and triclosan. That could happen if the genes in microorganisms that produce the enzyme ENR mutate, that is, change their chemical structure. Bacteria carrying the new gene might be more resistant to antimicrobial compounds.

Triclocarban has been shown to cause contact dermatitis, an allergic skin rash caused by contact with an irritating substance. This effect is of particular concern among infants and small children. The compound has been implicated in illnesses among newborns, and its use in maternity wards is discouraged. However, the compound is effective in treating other forms of skin rashes, such as atopic dermatitis, also known as eczema, because it kills the bacteria that cause the condition. Although the effects of ingesting triclocarban have not been thoroughly investigated, there is some question about the compound's possible carcinogenic effects.

Words to Know

ANTIMICROBIAL A substance that destroys microorganisms such as bacteria and fungi.

CARCINOGEN A chemical that causes cancer in humans or other animals.

ENZYME A complex protein that is produced by living cells and which promotes biochemical reactions in the body.

Triclocarban is considered by some authorities to be an environmental hazard, as well as a risk to human health. The compound is released into waterways when it is used for washing and bathing. Although wastewater treatment plants can remove up to 98 percent of the triclocarban in water, public health researchers have found the compound in 60 percent of the U.S. waterways studied. In one study, it was the fifth most common contaminant among 96 pollutants studied. The half life of triclocarban (the time it takes for half of the substance to disappear) is 1.5 years. That means that the compound will remain in water supplies for relatively long periods of time, certainly more than a few years. Although triclocarban is harmless to humans, it can be toxic to some types of aquatic life such as shellfish.

FOR FURTHER INFORMATION

"3,4,4'-Trichlorocarbanilide." Hazardous Substances Data Bank. http://toxnet.nlm.nih.gov/cgi-bin/sis/search (accessed on November 19, 2005).

"Anti-bacterial Additive Triclocarban Widespread in U.S. Waterways." Johns Hopkins Bloomberg School of Public Health, Public Health News Center. http://www.jhsph.edu/PublicHealthNews/Press_Releases/2005/Halden_triclocarban_triclosan.html (posted on January 21, 2005; accessed January 6, 2006).

Senese, Fred. "What Are Triclocarban and Triclosan (Ingredients in Some Antiseptic Soaps)?" General Chemistry Online! http://antoine.frostburg.edu/chem/senese/101/consumer/faq/triclosan.shtml (accessed on November 19, 2005).

See Also Triclosan

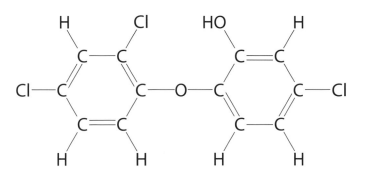

Triclosan

OVERVIEW

Triclosan (TRY-klo-san) is a white crystalline powder with antimicrobial properties that make it a useful ingredient in soaps, cosmetics, acne medications, deodorants, foot sprays and foot powders, toothpastes, and mouthwashes. It acts as an antimicrobial by inhibiting the action of an enzyme called enoyl-acyl carrier-protein reductase (ENR) that bacteria and fungi need to survive. The enzyme is used in the synthesis of fatty acids from which cell membranes are constructed. Having lost the ability to manufacture cell walls, bacteria and fungi die. The ENR enzyme is not present in humans, so triclosan has no effect on the human body.

Triclosan is also known by the following names: 5-chloro-2-(2,4-dichlorophenoxy)phenol; trichloro-2'-hydroxydiphenyl-ether; and 2,4,4'-trichloro-2'-hydroxydiphenyl ether.

HOW IT IS MADE

Two procedures are available for the synthesis of triclo-san. In one, 1,4-dichloro-2-nitrobenzene ($C_6H_3Cl_2NO_3$) is

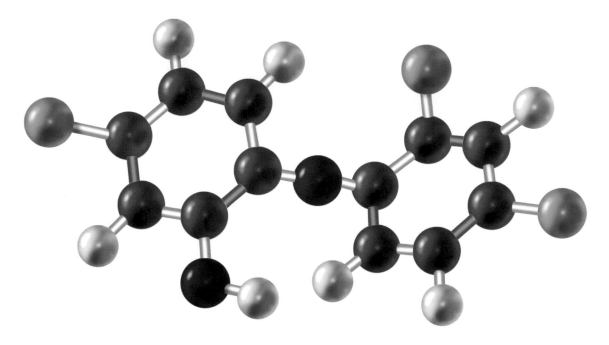

reacted with 2,4-dichlorophenol (C_6H_3ClOH) to obtain triclosan. In the second procedure, 2,4-dichlorobenzene ($C_6H_4Cl_2$), acetyl chloride (CH_3OCl), and 2,4-dichlorophenol (C_6H_3ClOH) are reacted with each other to obtain the product.

COMMON USES AND POTENTIAL HAZARDS

Triclosan has been used in soaps and other cleaning products since the 1960s. Hospitals found the compound to be very effective in controlling bacteria on hands and surfaces. Medical staff used it to stop outbreaks of infections caused by common bacteria such as *Staphylococcus aureus.* The compound is still used widely in hospitals for this purpose. In the 1990s, triclosan began to appear in a much wider array of products including antibacterial soaps, body washes, toothpastes, and dishwashing liquids. By the early 2000s, it was also being used in toothbrushes, plastic toys, and socks. The addition of triclosan to tooth care products is based on evidence that the compound may inhibit the growth of bacteria that cause dental caries (cavities). The compound has also proved to be effective in killing the

Interesting Facts

- Soaking socks in a solution that contains triclosan can prevent food odor because the compound kills the bacteria that grow on feet and are responsible for the odor.

parasites that cause malaria, one of the most serious diseases in the world.

In spite of its wide use, a number of questions have been raised about the inclusion of triclosan in household and commercial products. One objection is based on the concern that widespread use of the compound may lead to the development of bacteria resistant to antibiotics. Questions have also been raised about byproducts of the reactions by which triclosan is produced and by which it is degraded in the soil. Among those byproducts are a family of organic compounds known as the dioxins, among the most potent toxins known to humans. In addition, studies suggest that cleaning products that contain triclosan are not any more effective in killing germs than is the ordinary procedure of washing one's hands with soap and water. As a result, the Federal Trade Commission has ordered some companies to stop claiming that their product kills germs and reduces disease. The U.S. Environmental Protection Agency has listed triclosan as a possible human carcinogen.

Words to Know

CARCINOGEN A chemical that causes cancer in humans or other animals.

SYNTHESIS A chemical reaction in which some desired product is made from simple beginning chemicals, or reactants.

FOR FURTHER INFORMATION

Glaser, Aviva. "The Ubiquitous Triclosan." Beyond Pesticides. http://www.beyondpesticides.org/pesticides/factsheets/ Triclosan%20cited.pdf (accessed on November 19, 2005).

"Material Safety Data Sheet." Jeen International Corporation. http://www.jeen.com/cartexe/pdfs/msds%20JEECHEM%20 TRICLOSAN.pdf (accessed on November 19, 2005).

Mirsky, Steve. "Home, Bacteria-Ridden Home: Could Antibacterial Soaps Lead to Resistant Strains?" *Scientific American* (July 19, 1997). Also available online at http://sciam.com/article.cfm?articleID=0009078D-A418-1C76-9B81809EC588EF21 (accessed on November 19, 2005).

See Also Triclocarban

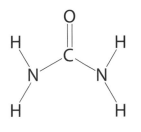

Urea

OVERVIEW

Urea (yoo-REE-uh) is a white crystalline solid or powder with almost no odor and a salty taste. It is a product of the decomposition of proteins in the bodies of terrestrial animals. Urea is produced in the liver and transferred to the kidneys, from which it is excreted in urine. The compound was first identified as a component of urine by French chemist Hilaire Marin Rouelle (1718-1799) in 1773. It was first synthesized accidentally in 1828 by German chemist Friedrich Wöhler (1800-1882). The synthesis of urea was one of the most important historical events in the history of chemistry. It was the first time that a scientist had synthesized an organic compound. Prior to Wöhler's discovery, scientists believed that organic compounds could be made only by the intervention of some supernatural force. Wöhler's discovery showed that organic compounds were subject to the same set of natural laws as were inorganic compounds (compounds for non-living substances). For this reason, Wöhler is often called the Father of Organic Chemistry.

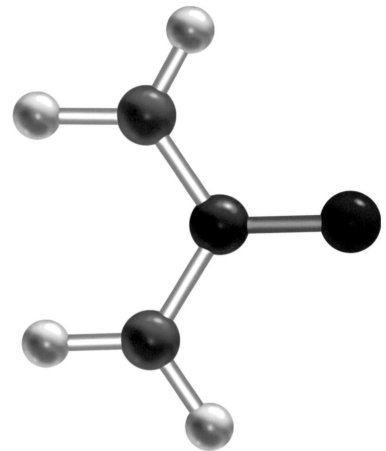

Urea. Red atom is oxygen; white atoms are hydrogen; black atom is carbon; and blue atoms are nitrogen. Gray sticks indicate double bonds. PUBLISHERS RESOURCE GROUP

HOW IT IS MADE

The formation of urea is the evolutionary solution to the problem of what to do with poisonous nitrogen compounds that formed when proteins decompose in the body. Proteins are large, complex compounds that contain relatively large amounts of nitrogen. When they decompose, that nitrogen is converted to ammonia (NH_3), a substance that is toxic to animals. If animals are to survive the decomposition of proteins (as happens whenever foods are metabolized), some method must be found to avoid the buildup of ammonia in the body.

That method involves a series of seven chemical reactions called the urea cycle by which nitrogen from proteins

CHEMICAL COMPOUNDS

Interesting Facts

- Different species of animals have evolved different methods of eliminating ammonia from their bodies. For example, fish excrete the ammonia produced by the decomposition of proteins directly into the watery environment in which they live. Birds, who consume less water by gram of weight than do most other animals, convert ammonia to uric acid ($C_5H_4N_4O_3$), a white crystalline solid that is even less toxic than urea.

- Urea is transported from the liver to the kidneys in the bloodstream. By the time it leaves the body in urine, its concentration is sixty to seventy times its concentration in the bloodstream.

is converted into urea. Although high concentrations of urea do pose a risk to animal bodies, the urea formed in these reactions is normally excreted fast enough to avoid health problems for an animal.

Urea is produced commercially by the direct synthesis of liquid ammonia (NH_3) and liquid carbon dioxide (CO_2). The product of this reaction is ammonium carbamate ($NH_4CO_2NH_2$):

$$2NH_3 + CO_2 \rightarrow NH_4CO_2NH_2$$

Ammonia and carbon dioxide do not react with each other under normal conditions of temperature and pressure. If the pressure is raised to 100 to 200 atmospheres (1750 to 3000 pounds per square inch) and the temperature is raised to about 200°C (400°C), however, the reaction proceeds efficiently with the formation of ammonium carbamate. When the pressure is then reduced to about 5 atmosphere (80 pounds per square inch), the ammonium carbamate decomposes to form urea and water:

$$NH_4CO_2NH_2 \rightarrow (NH_2)_2CO + H_2O$$

Words to Know

METABOLISM All of the chemical reactions that occur in cells by which fats, carbohydrates, and other compounds are broken down to produce energy and the compounds needed to build new cells and tissues.

SYNTHESIS A chemical reaction in which some desired chemical product is made from simple beginning chemicals, or reactants.

COMMON USES AND POTENTIAL HAZARDS

Urea is the sixteenth most important chemical in the United States, based on the amount produced annually. In 2004, the chemical industry produced 5.755 million metric tons (6.344 million short tons) of urea. Almost 90 percent of that output was used in the manufacture of fertilizers. An additional 5 percent went to the production of animal feeds. In both fertilizers and animal feeds, urea and the compounds from which it is made provide the nitrogen needed by growing plants and animals for their good health and survival. The other major use of urea is in the manufacture of various types of plastics, especially urea-formaldehyde resins and melamine.

Urea is also used:

- In the production of personal care products, such as hair conditioners, body lotions, and dental products;
- In certain pharmaceutical and medical products, such as creams to treat wounds and damaged skin;
- As a stabilizer in explosives, a compound that places limits on the rate at which an explosion proceeds;
- In the manufacture of adhesives;
- For the flame-proofing of fabrics;
- For the separation of products produced during the refining of petroleum;
- In the production of sulfamic acid ($HOSO_2NH_2$), an important raw material in many chemical processes;
- As a coating for paper products; and
- In the production of deicing agents.

FOR FURTHER INFORMATION

National Urea Cycle Disorders Foundation.
    http://www.nucdf.org/ (accessed on November 19, 2005).

Ophardt, Charles E. "Urea Cycle." Virtual Chembook.
    http://www.elmhurst.edu/~chm/vchembook/633ureacycle.
    html (accessed on November 19, 2005).

"Urea." International Programme on Chemical Safety.
    http://www.inchem.org/documents/icsc/icsc/eics0595.htm
    (accessed on November 19, 2005).

"Urea." Third Millennium Online.
    http://www.3rd1000.com/urea/urea.htm (accessed on November
    19, 2005).

See Also Ammonia

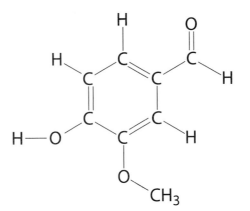

OTHER NAMES:
4-hydroxy-3-methoxy-
benzaldehyde;
3-methoxy-4-hydroxy-
benzaldehyde; vanillic
aldehyde

FORMULA:
$(CH_3O)(OH)C_6H_3CHO$

ELEMENTS:
Carbon, hydrogen,
oxygen

COMPOUND TYPE:
Ether (organic)

STATE:
Solid

MOLECULAR WEIGHT:
152.15 g/mol

MELTING POINT:
81.5°C (179°F)

BOILING POINT:
285°C (545°F)

SOLUBILITY:
Slightly soluble in
water; soluble in
glycerol, ethyl
alcohol, ether, and
acetone

KEY FACTS

Vanillin

OVERVIEW

Vanillin is a white crystalline solid with a pleasant, sweet aroma, and a characteristic vanilla-like flavor. Chemically, it is the methyl ether of 4-hydroxybenzoic acid, a ring compound that contains the carboxyl (-COOH) group and the hydroxyl (-OH) group. Vanillin is the substance responsible for the familiar taste of vanilla, which has been used as a food additive and spice for hundreds of years. Vanilla was probably first used as a flavoring by the inhabitants of South and Central America before the arrival of Europeans in the sixteenth century. Spanish explorers brought the spice back to Europe, where it soon became very popular as a food additive and for the flavoring of foods. Since that time, vanilla has become one of the world's most popular spices.

HOW IT IS MADE

Vanilla is obtained naturally from the seed pod of the tropical orchid *Vanilla planifolia* by a lengthy and expensive

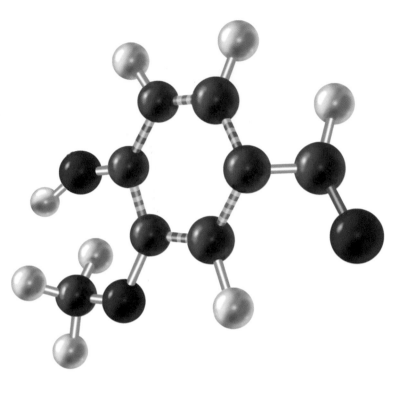

process. The pods are picked before they ripen and then cured until they are dark brown. The curing process involves soaking the pods in hot water, sun-drying them, and allowing them to "sweat" in straw. The cured pods are then soaked in alcohol to produce a product known as pure vanilla extract. The primary constituent in pure vanilla extract is vanillin, which gives the product its flavor. The process of extracting pure vanilla from seed pods may take as long as nine months.

Some people prefer a vanilla product that contains no, or almost no, alcohol. If alcohol is removed, almost pure vanilla is left behind, leaving a product known as natural vanilla flavoring.

Vanilla can also be extracted from plants other than *Vanilla planifolia,* such as potato peels and pine tree sap. The most economical source of the product, however, is waste material left over from the wood pulp industry. That waste material consists primarily of lignin, a complex natural polymer that, along with cellulose, is the primary component of wood. The wastes from wood pulping can be treated to break down and separate the lignin. This leaves behind a complex mixture, a major component of which is vanilla. This vanilla is called lignin vanilla and has many of the same physical properties as natural vanilla. Since it is so much less expensive to make, it has become one of the major forms of vanilla used by consumers. Lignin vanilla is known commercially as artificial vanilla flavoring.

The two forms of vanilla described earlier—natural vanilla and lignin vanilla—are mixtures in which the compound vanillin is a major component. In both mixtures, other components are present in lesser amounts. These components may add somewhat different flavors and aromas, modifying the pure taste and smell of vanillin. Artificial methods for the production of pure vanillin have been available since the late 1890s. The most popular of those methods begins with eugenol ($(C_3H_5)C_6H_3(OH)OCH_3$) or isoeugenol ($(CH_3CHCH)C_6H_3(OH)OCH_3$). Either of these compounds is then treated with acetic anhydride ($(CH_3CO)_2O$) to obtain vanillin acetate, which is then converted to pure vanillin. The product of this reaction, unlike natural vanilla and lignin vanilla, is a pure compound, 4-hydroxy-3-methoxybenzaldehyde, pure vanillin. This method was the primary method for making artificial vanillin for more than 50 years. It has since been replaced by an alternative method of preparation, the Reimer-Tiemann reaction. This method for making artificial vanillin begins with catechol ($C_6H_4(OH)_2$) or guaiacol ($CH_3OC_6H_4OH$).

The Remier-Tiemann reaction is also used to produce another form of vanillin called ethyl vanillin. Ethyl vanillin is the ethyl ether of 4-hydroxybenzoic acid, 4-hydroxy-3-ethoxybenzaldehyde ($(CH_3CH_2O)(OH)C_6H_3CHO$). It is a close chemical relative of natural vanillin in which the methyl (-CH_3) group of natural vanillin is replaced by an ethyl (-CH_2CH_3) group. Ethyl vanillin is also known as artificial vanilla or synthetic vanilla. Its flavor is about three times as strong as that of methyl vanillin and is used to fortify or replace natural vanillin and lignin vanillin.

COMMON USES AND POTENTIAL HAZARDS

All forms of vanillin are used as a flavoring agent and sweetener in many types of foods, including candies, dessert products, ice creams, puddings, yogurts, diet shakes, and soft drinks. It is also added to some wines, alcoholic liquors,

Interesting Facts

- The tropical orchid *Vanilla planifolia* is pollinated naturally by the tiny Melipone bee, native to Mexico. In areas where the bee does not live, the orchid must be pollinated artificially by humans.

- Less than 1 percent of the vanillin produced annually comes from vanilla beans. The remaining 99 percent comes from lignin or is produced by synthetic means.

- Small amounts of vanillin are present in the wood used to make wine casks and adds to the flavor of wine.

- The Coca-Cola Company is believed to be the world's largest buyer of pure vanillin. Since the company does not reveal the recipe for its products, that assumption cannot be confirmed.

- The word *vanilla* comes from the Spanish word *vainilla*, meaning "little sheath," which refers to the shape of the vanilla orchid.

toothpastes, and cigarettes. The vanillins have also been shown to stimulate one's appetite, so they have been used to treat appetite loss. They are also added to cattle feed to enhance weight gain.

However, less than half the vanillin produced is used in food products. Vanillin's rich fragrance makes the compound useful also as an additive in perfumes, air fresheners, soaps, shampoos, candles, creams, lotions, colognes, and ointments. The compound is also used as a raw material in the manufacture of a variety of drugs, particularly the compound known as L-dopa, used to treat Parkinson's disease.

Vanillin is considered safe for human consumption, although it can be toxic in very large quantities. Known reactions include respiratory irritation, including coughing and shortness of breath, and gastrointestinal tract irritation. Contact with the skin or eyes can also cause irritation, redness, and pain. These symptoms are virtually unknown except for individuals who work directly with the pure compounds.

FOR FURTHER INFORMATION

"All about Vanilla Extracts and Flavors." The Vanilla.COMpany. http://www.vanilla.com/html/facts-extracts.html (accessed on November 19, 2005).

"Aroma Chemicals from Petrochemical Feedstocks." National Economic and Development Council of South Africa. http://www.nedlac.org.za/research/fridge/aroma/part3/benchmarking.pdf (accessed on November 19, 2005).

"Food Guide Question." Gloucester Muslim Welfare Association. http://www.gmwa.org.uk/foodguide2/viewquestion.php?foodqid=69&catID=1&compID=1 (accessed on November 19, 2005).

Rain, Patricia. *Vanilla: A Cultural History of the World's Most Popular Flavor and Fragrance.* Edited by Jeremy P. Tarcher. New York: Penguin Group USA, 2004.

"Vanillin." Greener Industry. http://www.uyseg.org/greener_industry/pages/vanillin/1Vanillin_AP.htm (accessed on November 19, 2005).

Water. Red atom is oxygen and
white atoms are hydrogen.
PUBLISHERS RESOURCE GROUP

For example, water has a very high boiling point for a substance with relatively small molecules. The high boiling point is a result of the fact that heat added to water must first be used to break hydrogen bonds between water molecules before providing enough energy to vaporize the molecules. Similarly, the phenomenon known as surface tension is caused by hydrogen bonding. Surface tension is the tendency of a liquid to act as if it is covered with a thin film. Some insects are able to walk on water because its surface tension is so great. The surface tension is caused by the attractive forces between adjacent water molecules.

Water is also an excellent solvent. A solvent is a substance capable of dissolving other substances. Chemists sometimes refer to water as "the universal solvent" because it is able to dissolve so many other substances. That statement is an exaggeration, but does reflect the compound's ability to dissolve more substances that probably any other single compound. Water's ability to dissolve other substances is at least partly a result of its strong dipole character. The positive or negative end of a water molecule attaches itself to the negative or positive end of the substance to be dissolved. The force of attraction exerted by the water molecule is sufficient to tear apart the particles of which the second substance is composed causing them to dissolve in the water.

Interesting Facts

- When water freezes, the attractive forces between molecules force them into a regular crystalline pattern that occupies more space than do the same water molecules in the liquid state. In other words, water expands when it freezes and ice is less dense than liquid water.

- One consequence of that fact is that lakes freeze from the top down. As ice forms, it floats to the surface of the liquid. Aquatic organisms in the lake are able to live through the winter because of the layer of ice on top of the lake.

- Adding to the protection added by the layer of ice is the fact that ice is one of the best thermal (heat) insulators known. That is, heat flows through ice more slowly than through almost any other substance. Any heat left in a freezing lake will be conserved within the water since it escapes so slowly through the ice layer.

HOW IT IS MADE

Water can be made by a variety of chemical reactions, including:

- The oxidation of hydrogen: $2H_2 + O_2 \rightarrow 2H_2O$;

- The reaction between an acid and a base, as, for example: $NaOH + HCl \rightarrow NaCl + H_2O$;

- The combustion of an organic material, as, for example: $CH_4 + 2O_2 \rightarrow CO_2 + 2H_2O$.

Because water occurs so abundantly, none of these reactions is required for the commercial production of the compound. Water makes up about 70 percent of the Earth's surface in the oceans, lakes, rivers, ponds, glaciers, ice caps, and other reservoirs. The problem is that only a very small fraction of that water—about 3 percent—is fresh water. The remaining 97 percent is salt water. And even the 3 percent of fresh water in lakes, rivers, and other resources is impure, in the sense that it contains other substances dissolved and suspended in it.

Thus, the primary concern in obtaining adequate supplies of pure water for household, personal, commercial, industrial, or other uses is the purification of water, not its synthesis. Purification of water is achieved by a number of processes, including chlorination, filtration, distillation, or purification by some type of ion exchange mechanism.

COMMON USES AND POTENTIAL HAZARDS

One of the most important uses of water is the survival of life on Earth. All plants and animals contain a high proportion of water. That proportion ranges from as high as 97 percent in many fruits and vegetables to a low of about 20 percent in some "dry" foods like breads and cereals. The human body itself comprises anywhere from 50 to 70 percent of a person's body weight. That water plays a number of roles, such as providing a solvent by which nutrients are circulated throughout the body and making possible all kinds of chemical reactions that occur in aqueous solutions, but do not occur in the dry state.

Although easy to ignore, water plays an almost unlimited number of roles in industrial, chemical, commercial, and other operations. Among these applications of water are:

- As a coolant in electricity generating plants, petroleum refineries, chemical plants, and many kinds of industrial factories;
- For irrigation;
- For cleaning, washing, and scouring raw materials and finished products;
- As a source of hydrogen and oxygen for many industrial and chemical operations;
- As a solvent for many types of industrial and chemical reactions and for the extraction of compounds from mixtures;
- In the manufacture of many kinds of foods and beverages, such as beer, wine, and soft drinks;
- As a medium for suspending materials in industrial processes, such as the manufacture of paper;
- In the processing of textiles;
- For the generation of steam to power industrial, household, and chemical processes;

Words to Know

AQUEOUS SOLUTION A solution that consists of some material dissolved in water.

ELECTROLYTE A substance which, when dissolved in water, will conduct an electric current.

HYDROLYSIS The process by which a compound reacts with water to form two new compounds.

- For the hydration of lime;
- As a coolant in nuclear reactors;
- In the transport of industrial and chemical raw materials and products;
- For the removal of barks from logs in the timber industry;
- To make possible hydrolysis reactions in chemical and industrial operations;
- In the manufacture of Portland cement; and
- To dilute solutions that are too concentrated for some given industrial process.

FOR FURTHER INFORMATION

"Iowa Project WET." Iowa Academy of Science. http://www.uni.edu/~iowawet/iowawet.html (accessed on November 19, 2005).

Strange, Veronica. *The Meaning of Water.* Oxford, UK: Berg Publishers, 2004.

"Water Properties." London South Bank University. http://www.lsbu.ac.uk/water/data.html (accessed on November 19, 2005).

"Water Science for Schools." U.S. Geological Survey. http://ga.water.usgs.gov/edu/ (accessed on November 19, 2005).

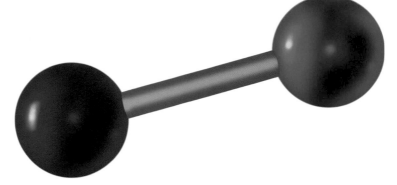

Zinc oxide. Red atoms are oxygen and turquoise atoms are zinc. Gray sticks indicate double bonds. PUBLISHERS RESOURCE GROUP

carbonate ($ZnCO_3 \cdot 2Zn(OH)_2$). Leaching is a process by which soluble salts are dissolved out of soil, ore, or some other material. The basic zinc carbonate is decomposed by heat, yielding pure zinc oxide:

$$ZnCO_3 \cdot 2Zn(OH)_2 \rightarrow 3ZnO + CO_2 + 2H_2O$$

COMMON USES AND POTENTIAL HAZARDS

One of zinc oxide's most important uses historically is in paints and pigments. Common names by which it is now known, such as Chinese white and zinc white, are terms long used by artists for the compound. White paints and pigments made with zinc oxide retain their luster and purity longer than most other types of white paints, partly because zinc oxide does not react readily with hydrogen sulfide in air that causes most white paints to darken.

One of zinc oxide's most important properties is its ability to absorb ultraviolet (UV) light in sunlight. Because of this property, zinc oxide is often added to sunscreens and sunblocks to help protect a person from sunburn. The same property accounts for other important applications for zinc oxide, such as their use in rubber and plastic products. By absorbing ultraviolet light, zinc oxide protects the rubber or plastic from decomposing. The compound also has other applications in the rubber and plastic industries, including:

• In the curing of natural and synthetic rubber products, increasing the rate at which the product reaches its final chemical state;

Interesting Facts

- The term *philosopher's wool* dates to the period of alchemy when scholars often gave picturesque and descriptive names to the substances with which they worked. Philosopher's wool referred to the fluffy white material that was formed when alchemists (also known as natural philosophers) heated zinc in air. The substance was also referred to as *nix alba,* or "white snow" for the same reason.

- As a fungicidal additive to rubber and plastic, preventing fungi from attacking and destroying products made from those materials;

- To increase the temperature at which rubber and plastic products remain stable and the amount of exposure to light they can withstand;

- To maintain the proper acidic properties of a product, reducing the rate at which the are likely to decay;

- To provide additional strength to the rubber or plastic product.

Besides its major uses in the rubber and plastic industries, zinc oxide has a number of applications in other fields, such as:

- As an additive in glass and ceramic materials, to provide greater heat resistance, greater resistance to breakage by shock, and high luster;

- In the manufacture of specialized types of sealants and adhesives;

- As a light-gathering agent in photocopy machines;

- As an additive in lubricants for the purpose of reducing wear and helping the lubricant withstand high pressures;

- In smokestacks as an aid in removing sulfur dioxide and other pollutant gases produced in factory operations; and

Words to Know

ALCHEMY An ancient field of study from which the modern science of chemistry evolved.

AQUEOUS SOLUTION A solution that consists of some material dissolved in water.

• In the manufacture of specialized packaging materials, especially those used for food products, because of the compound's ability to kill certain microorganisms that cause food decay.

In moderate amounts, zinc oxide is a relatively harmless compound. Exposure to zinc oxide dust may cause respiratory problems, such as coughing, upper respiratory tract irritation, chills, fever, nausea, and vomiting.

FOR FURTHER INFORMATION

"Application." Nav Bahrat Metallic Oxides Industries. http://www.navbharat.co.in/clients.htm (accessed on November 19, 2005).

"Occupational Safety and Health Guideline for Zinc Oxide." Occupational Safety & Health Administration. http://www.osha.gov/SLTC/healthguidelines/zincoxide/recognition.html (accessed on November 19, 2005).

"Zinc Oxide." Center for Advanced Microstructures and Devices, Louisiana State University. http://www.camd.lsu.edu/msds/z/zinc_oxide.htm (accessed on November 19, 2005).

"Zinc Oxide Producers Association." http://www.cefic.be/Templates/shwAssocDetails.asp?NID=5&HID=25&ID=172 (accessed on November 19, 2005).

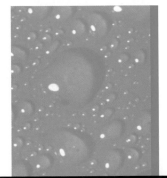

appendices

Lists of compounds

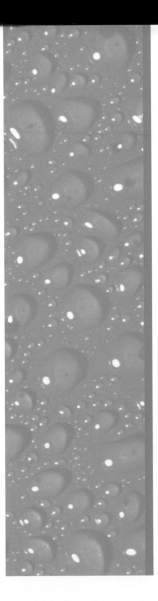

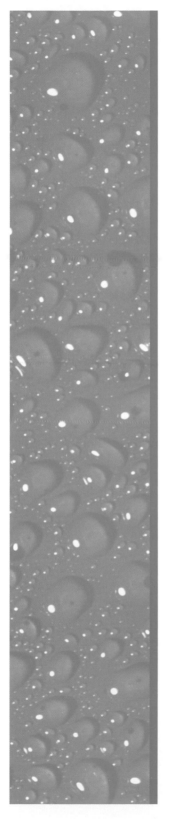

Compounds by Formula

AgI
  Silver Iodide

AgNO$_3$
  Silver Nitrate

Ag$_2$O
  Silver(I) Oxide

Ag$_2$S
  Silver(I) Sulfide

AlF$_3$
  Aluminum Fluoride

Al(OH)$_3$
  Aluminum Hydroxide

Al$_2$O$_3$
  Aluminum Oxide

CCl$_2$F$_2$
  Dichlorodifluoromethane

CCl$_4$
  Carbon Tetrachloride

-[-CF$_2$-]-$_n$
  Polytetrafluoroethylene

CH≡CH
  Acetylene

-[-CH(CH$_3$)CH$_2$-]-$_n$
  Polypropylene

CHCl$_3$
  Chloroform

CH$_2$O
  Formaldehyde

-[-CH$_2$C(CH$_3$)(COOH)CH$_2$-]-$_n$
  Polymethyl Methacrylate

CH$_2$=C(CN)COOCH$_3$
  Cyanoacrylate

CH$_2$=CHCH-CH$_2$
  1,3-Butadiene

CH$_2$=CHCH$_3$
  Propylene

CH$_2$=CH(CH$_3$)CH=CH$_2$
  Isoprene

-[-CH$_2$-CH(COONa)-]-$_n$-
  Sodium Polyacrylate

-[CH$_2$CHC$_6$H$_5$-]-$_n$-
[-CH$_2$CH=CHCH$_2$-]-$_n$-[CH$_2$CHC$_6$H
  Poly(Styrene-Butadiene-
  Styrene)

-[-CH$_2$CHCl-]-$_n$
  Polyvinyl Chloride

CH$_2$=CH$_2$
  Ethylene

(CH$_2$CH$_2$Cl)$_2$S
  2,2′-Dichlorodiethyl Sulfide

-[-CH_2-CH_2-]-$_n$
Polyethylene

-[-$CH_2C_6H_5$-]-$_n$
Polystyrene

$CH_2NO_2CHNOCH_2NO_2$
Nitroglycerin

$CH_2OHCHOHCH_2OH$
Glycerol

CH_2OHCH_2OH
Ethylene Glycol

$(CH_2)_2O$
Ethylene Oxide

$CH_3CHOHCH_3$
Isopropyl Alcohol

$CH_3CHOHCOOH$
Lactic Acid

CH_3CH_2OH
Ethyl Alcohol

CH_3COCH_3
Dimethyl Ketone

$CH_3CONHC_6H_4OH$
Acetaminophen

$CH_3COOCH_2CH_2CH(CH_3)_2$
Isoamyl Acetate

$CH_3COOC_2H_5$
Ethyl Acetate

$CH_3COOC_5H_{11}$
Amyl Acetate

$CH_3COOC_6H_4COOH$
Acetylsalicylic acid

CH_3COOH
Acetic acid

$CH_3C_5HN(OH)(CH_2OH)_2$
Pyridoxine

$CH_3C_6H_9$ $(C_3H_7)OH$
Menthol

CH_3OH
Methyl Alcohol

$(CH_3O)(OH)C_6H_3CHO$
Vanillin

CH_3SH
Methyl Mercaptan

$(CH_3)_2CHCH_2CH_2NO_2$
Amyl Nitrite

$C_5H_{11}NO_2$
Amyl Nitrite

$(CH_3)_2C_5H_3NSO(COOH)NHCOR$
Penicillin

$(CH_3)_3COCH_3$
Methyl-t-butyl Ether

CH_4
Methane

CH_4S
Methyl Mercaptan

CO
Carbon Monoxide

-[-$CO(CH_2)_4CO$-$NH(CH_2)_6NH$-]-$_n$
Nylon 6 and Nylon 66

-[-$CO(CH_2)_5NH$-]-$_n$
Nylon 6 and Nylon 66

-[-$CONH$-C_6H_4-$NCOO$-CH_2CH_2-O-]-$_n$
Polyurethane

$COOH(CH_2)_2CH(NH_2)COONa$
Monosodium Glutamate

CO_2
Carbon Dioxide

C_2H_2
Acetylene

$C_2H_2O_4$
Oxalic Acid

$[C_2H_3Cl]_n$
Polyvinyl Chloride

C_2H_4
Ethylene

$[C_2H_4]_n$
Polyethylene

C_2H_4O
Ethylene Oxide

$C_2H_4O_2$
Acetic acid

C_2H_6O
Ethyl Alcohol

$C_2H_6O_2$
Ethylene Glycol

$[C_3H_3O_2Na]_n$
Sodium Polyacrylate

$C_3H_5N_3O_5$
Nitroglycerin

$[C_3H_6]_n$
Polypropylene

C_3H_6O
Dimethyl Ketone

$C_3H_6O_3$
Lactic Acid

C_3H_8
Propane

C_3H_8O
Isopropyl Alcohol

$C_3H_8O_3$
Glycerol

C_4H_6
1,3-Butadiene

$C_4H_8Cl_2S$
2,2'-Dichlorodiethyl Sulfide

$C_4H_8O_2$
Ethyl Acetate

C_4H_{10}
Butane

$C_4H_{10}S$
Butyl Mercaptan

$C_5H_4NC_4H_7NCH_3$
Nicotine

$C_5H_5NO_2$
Cyanoacrylate

C_5H_8
Isoprene

$[C_5H_8]_n$
Polymethyl Methacrylate

$C_5H_8NNaO_4$
Monosodium Glutamate

$C_5H_{12}O$
Methyl-t-butyl Ether

$C_6H_2(CH_3)(NO_2)_3$
2,4,6-Trinitrotoluene

$C_6H_3ClOH-O-C_6H_3Cl_2$
Triclosan

$C_6H_3Cl_2NHCONHC_6H_4Cl$
Triclocarban

$C_6H_5CH=CHCHO$
Cinnamaldehyde

$C_6H_5CH=CH_2$
Styrene

$C_6H_5CH(CH_3)_2$
Cumene

$C_6H_5CH_3$
Toluene

C_6H_5COOH
Benzoic Acid

$C_6H_5C_2H_5$
Ethylbenzene

$C_6H_5NO_2$
Niacin

C_6H_5OH
Phenol

C_6H_6
Benzene

$C_6H_6Cl_6$
Gamma-1,2,3,4,5,6-Hexa-
chlorocyclohexane

C_6H_6O
Phenol

$[C_6H_7O_2(OH)_2OCS_2Na]_n$
Cellulose Xanthate

$C_6H_8O_6$
Ascorbic Acid

$C_6H_8O_7$
Citric Acid

$(C_6H_{10}O_5)_n$
Cellulose

$[C_6H_{11}NO]_n$
Nylon 6 and Nylon 66

$C_6H_{11}NHSO_3Na$
Sodium Cyclamate

$C_6H_{12}NNaSO_3$
Sodium Cyclamate

$C_6H_{12}O_2$
Butyl Acetate

$C_6H_{12}O_6$
Fructose
Glucose

C_6H_{14}
Hexane

$C_6H_4(CH_3)CON(C_2H_5)_2$
N,N-Diethyl-3-Methylbenza-
mide

$C_7H_5NO_3S$
Saccharin

$C_7H_5N_3O_6$
2,4,6-Trinitrotoluene

$C_7H_6O_2$
Benzoic Acid

$[C_7H_7]_n$
Polystyrene

$C_8H_7N_3O_2$
Luminol

$C_7H_8N_4O_2$
Theobromine

$[C_7H_{11}NaO_5S_2]_n$
Chloroform

$C_7H_{14}O_2$
Isoamyl Acetate
Amyl Acetate

C_8H_8
Styrene

$[C_8H_8]_n-[C_4H_6]_n-[C_8H_8]_n$
Poly(Styrene-Butadiene-
Styrene)

$C_8H_9NO_2$
Acetaminophen

C_8H_{10}
Ethylbenzene

$C_8H_{10}N_4O_2$
Caffeine

$C_8H_{11}NO_3$
Pyridoxine

C_9H_8O
Cinnamaldehyde

$C_9H_8O_4$
Acetylsalicylic acid

C_9H_{12}
Cumene

$C_{10}H_8$
Naphthalene

$C_{10}H_{14}N_2$
Nicotine

$C_{10}H_{16}O$
Camphor

$C_{10}H_{20}O$
Menthol

$C_{11}H_{16}O_2$
Butylated Hydroxyanisole
and Butylated Hydroxyto-
luene (BHA)

$C_{12}H_7Cl_3O_2$
Triclosan

$C_{12}H_{16}N_4O_{18}$
Cellulose Nitrate

$C_{12}H_{17}ClN_4OS$
Thiamine

$C_{12}H_{17}NO$
N,N-Diethyl-3-Methylbenza-
mide

$[C_{12}H_{22}N_2O_2]_n$
Nylon 6 and Nylon 66

$C_{12}H_{22}O_{11}$
Lactose
Sucrose

$C_{13}H_9Cl_3N_2O$
Triclocarban

$C_{13}H_{18}O_2$
2-(4-Isobutylphenyl)propionic
Acid

$C_{14}H_9Cl_5$
Dichlorodiphenyltrichloro-
ethane

$C_{14}H_{14}O_3$
Naproxen

$C_{14}H_{18}N_2O_5$
L-Aspartyl-L-Phenylalanine
Methyl Ester

$C_{15}H_{24}O$
Butylated Hydroxyanisole
and Butylated Hydroxyto-
luene (BHT)

$C_{16}H_{18}N_2OS$
Penicillin

$C_{16}H_{19}N_3O_5S$
Amoxicillin

$C_{17}H_{20}N_4O_6$
Riboflavin

$C_{19}H_{19}N_7O_6$
Folic Acid

$C_{19}H_{28}O_2$
Testosterone

$C_{20}H_{30}O$
Retinol

$C_{27}H_{45}OH$
Cholesterol

$C_{28}H_{34}N_2O_3$
Denatonium Benzoate

$C_{29}H_{50}O$
Alpha-Tocopherol

$C_{35}H_{28}O_5N_4Mg$
Chlorophyll

$C_{35}H_{30}O_5N_4Mg$
Chlorophyll

$C_{40}H_{56}$
Beta-Carotene

$C_{54}H_{70}O_6N_4Mg$
Chlorophyll

$C_{55}H_{70}O_6N_4Mg$
Chlorophyll

$C_{55}H_{72}O_5N_4Mg$
Chlorophyll

$C_{63}H_{88}CoN_{14}O_{14}P$
Cyanocobalamin

$C_{76}H_{52}O_{46}$
Tannic Acid

$CaCO_3$
Calcium Carbonate

$CaHPO_4$
Calcium Phosphate

CaO
Calcium Oxide

$Ca(OH)_2$
Calcium Hydroxide

$CaSO_4$
Calcium Sulfate

$CaSiO_3$
Calcium Silicate

$Ca(H_2PO_4)_2$
Calcium Phosphate

$Ca_3(PO_4)_2$
Calcium Phosphate

$(ClC_6H_4)_2CHCCl_3$
Dichlorodiphenyltrichloro-
ethane

$-ClO_4$
Perchlorates

CuO
Copper(II) Oxide

$CuSO_4$
Copper(II) Sulfate

Cu_2O
Copper(I) Oxide

FeO
Iron(II) Oxide

Fe_2O_3
Iron(III) Oxide

$HCHO$
Formaldehyde

HCl
Hydrogen Chloride

HNO_3
Nitric Acid

HOH
Water

HOOCCH$_2$C(OH)(COOH)CH$_2$COOH
Citric Acid

HOOCCOOH
Oxalic Acid

H$_2$O
Water

H$_2$O$_2$
Hydrogen Peroxide

H$_2$SO$_4$
Sulfuric Acid

H$_3$BO$_3$
Boric Acid

H$_3$PO$_4$
Phosphoric Acid

HgS
Mercury(II) Sulfide

KAl(SO$_4$)$_2$
Aluminum Potassium Sulfate

KCl
Potassium Chloride

KF
Potassium Fluoride

KHCO$_3$
Potassium Bicarbonate

KHC$_4$H$_4$O$_6$
Potassium Bitartrate

KHSO$_4$
Potassium Bisulfate

KI
Potassium Iodide

KNO$_3$
Potassium Nitrate

KOH
Potassium Hydroxide

K$_2$CO$_3$ K$_{34}$NO$_{15}$
Potassium Carbonate

K$_2$SO$_4$
Potassium Sulfate

MgCl$_2$
Magnesium Chloride

MgO
Magnesium Oxide

Mg(OH)$_2$
Magnesium Hydroxide

MgSO$_4$
Magnesium Sulfate

Mg$_3$Si$_4$O$_{10}$(OH)$_2$
Magnesium Silicate Hydroxide

(NH$_2$)$_2$CO
Urea

NH$_3$
Ammonia

(NH$_4$)$_2$SO$_4$
Ammonium Sulfate

NH$_4$Cl
Ammonium Chloride

NH$_4$NO$_3$
Ammonium Nitrate

NH$_4$OH
Ammonium Hydroxide

NO
Nitric Oxide

NO$_2$
Nitrogen Dioxide

N$_2$O
Nitrous Oxide

NaBO$_3$
Sodium Perborate

NaC$_2$H$_3$O$_2$
Sodium Acetate

NaCl
Sodium Chloride

NaClO
Sodium Hypochlorite

NaF
Sodium Fluoride

NaHCO$_3$
Sodium Bicarbonate

NaH$_2$PO$_4$
Sodium Phosphate

NaOH
Sodium Hydroxide

Na$_2$B$_4$O$_7$
Sodium Tetraborate

Na$_2$B$_4$O$_7$·10H$_2$O
Sodium Tetraborate

Na$_2$CO$_3$
Sodium Carbonate

Na$_2$HPO$_4$
Sodium Phosphate

Na$_2$SO$_3$
Sodium Sulfite

Na$_2$S$_2$O$_3$
Sodium Thiosulfate

Na$_2$SiO$_3$
Sodium Silicate

Na$_3$PO$_4$
Sodium Phosphate

SO$_2$
Sulfur Dioxide

SiO$_2$
Silicon Dioxide

SnF$_2$
Stannous Fluoride

ZnO
Zinc Oxide

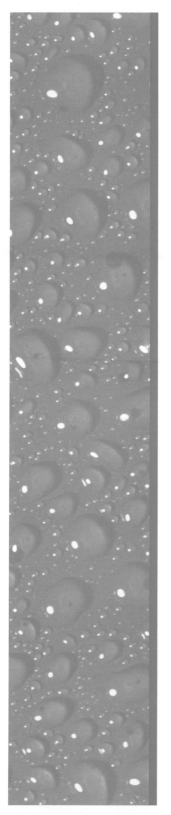

Compounds by Element

ALUMINUM

Aluminum Fluoride
Aluminum Hydroxide
Aluminum Oxide
Aluminum Potassium Sulfate

BORON

Boric Acid
Sodium Perborate
Sodium Tetraborate

CALCIUM

Calcium Carbonate
Calcium Hydroxide
Calcium Oxide
Calcium Phosphate
Calcium Silicate
Calcium Sulfate

CARBON

1,3-Butadiene
2-(4-Isobutylphenyl)propionic
    Acid
2,2$'$-Dichlorodiethyl
    Sulfide
2,4,6-Trinitrotoluene
Acetaminophen
Acetic acid
Acetylene
Acetylsalicylic acid
Alpha-Tocopherol

Amoxicillin
Amyl Acetate
Amyl Nitrite
Ascorbic Acid
Benzene
Benzoic Acid
Beta-Carotene
Butane
Butyl Acetate
Butyl Mercaptan
Butylated Hydroxyanisole
    and Butylated
    Hydroxytoluene
Caffeine
Calcium Carbonate
Camphor
Carbon Dioxide
Carbon Monoxide
Carbon Tetrachloride
Cellulose
Cellulose Nitrate
Cellulose Xanthate
Chloroform
Chlorophyll
Cholesterol
Cinnamaldehyde
Citric Acid
Collagen
Cumene
Cyanoacrylate
Cyanocobalamin
Denatonium Benzoate
Dichlorodifluoromethane

Dichlorodiphenyltrichloro-
    ethane
Dimethyl Ketone
Ethyl Acetate
Ethyl Alcohol
Ethylbenzene
Ethylene
Ethylene Glycol
Ethylene Oxide
Folic Acid
Formaldehyde
Fructose
Gamma-1,2,3,4,5,6-Hexachloro-
    cyclohexane
Gelatin
Glucose
Glycerol
Hexane
Isoamyl Acetate
Isoprene
Isopropyl Alcohol
Lactic Acid
Lactose
L-Aspartyl-L-Phenylalanine
    Methyl Ester
Luminol
Menthol
Methane
Methyl Alcohol
Methyl Mercaptan
Methyl-t-butyl Ether
Monosodium Glutamate
N,N-Diethyl-3-Methyl-
    benzamide
Naphthalene
Naproxen
Niacin
Nicotine
Nitroglycerin
Nylon 6 and Nylon 66
Oxalic Acid
Pectin
Penicillin
Petrolatum
Petroleum
Phenol
Poly(Styrene-Butadiene-
    Styrene)

Polycarbonates
Polyethylene
Polymethyl Methacrylate
Polypropylene
Polysiloxane
Polystyrene
Polytetrafluoroethylene
Polyurethane
Polyvinyl Chloride
Potassium Bicarbonate
Potassium Bitartrate
Potassium Carbonate
Propane
Propylene
Pyridoxine
Retinol
Riboflavin
Saccharin
Sodium Acetate
Sodium Bicarbonate
Sodium Carbonate
Sodium Cyclamate
Sodium Polyacrylate
Styrene
Sucrose
Sucrose Polyester
Tannic Acid
Testosterone
Theobromine
Thiamine
Toluene
Triclocarban
Triclosan
Urea
Vanillin

CHLORINE

2,2′-Dichlorodiethyl Sulfide
Ammonium Chloride
Carbon Tetrachloride
Chloroform
Dichlorodifluoromethane
Dichlorodiphenyltrichloro-
    ethane
Gamma-1,2,3,4,5,6-Hexachloro-
    cyclohexane
Hydrogen Chloride

Magnesium Chloride
Perchlorates
Polyvinyl Chloride
Potassium Chloride
Sodium Chloride
Sodium Hypochlorite
Thiamine
Triclocarban
Triclosan

COBALT

Cyanocobalamin

COPPER

Copper(I) Oxide
Copper(II) Oxide
Copper(II) Sulfate

FLUORINE

Aluminum Fluoride
Dichlorodifluoromethane
Polytetrafluoroethylene
Potassium Fluoride
Sodium Fluoride
Stannous Fluoride

HYDROGEN

1,3-Butadiene
2-(4-Isobutylphenyl)propionic
    Acid
2,2′-Dichlorodiethyl Sulfide
2,4,6-Trinitrotoluene
Acetaminophen
Acetic acid
Acetylene
Acetylsalicylic acid
Alpha-Tocopherol
Ammonia
Ammonium Chloride
Ammonium Hydroxide
Ammonium Nitrate
Ammonium Sulfate
Amoxicillin
Amyl Acetate
Amyl Nitrite
Ascorbic Acid

Benzene
Benzoic Acid
Beta-Carotene
Boric Acid
Butane
Butyl Acetate
Butyl Mercaptan
Butylated Hydroxyanisole and
    Butylated Hydroxytoluene
Caffeine
Calcium Hydroxide
Calcium Phosphate
Camphor
Cellulose
Cellulose Nitrate
Cellulose Xanthate
Chloroform
Chlorophyll
Cholesterol
Cinnamaldehyde
Citric Acid
Collagen
Cumene
Cyanoacrylate
Cyanocobalamin
Denatonium Benzoate
Dichlorodiphenyltrichloro-
    ethane
Dimethyl Ketone
Ethyl Acetate
Ethyl Alcohol
Ethylbenzene
Ethylene
Ethylene Glycol
Ethylene Oxide
Folic Acid
Formaldehyde
Fructose
Gamma-1,2,3,4,5,6-Hexachloro-
    cyclohexane
Gelatin
Glucose
Glycerol
Hexane
Hydrogen Chloride
Isoamyl Acetate
Isoprene
Isopropyl Alcohol

Lactic Acid
Lactose
L-Aspartyl-L-Phenylalanine
    Methyl Ester
Luminol
Magnesium Hydroxide
Magnesium Silicate
    Hydroxide
Menthol
Methane
Methyl Alcohol
Methyl Mercaptan
Methyl-t-butyl Ether
Monosodium Glutamate
N,N-Diethyl-3-Methyl-
    benzamide
Naphthalene
Naproxen
Niacin
Nicotine
Nitric Acid
Nitroglycerin
Nylon 6 and Nylon 66
Oxalic Acid
Pectin
Penicillin
Petrolatum
Petroleum
Phenol
Phosphoric Acid
Poly(Styrene-Butadiene-
    Styrene)
Polycarbonates
Polyethylene
Polymethyl
    Methacrylate
Polypropylene
Polysiloxane
Polystyrene
Polyurethane
Polyvinyl Chloride
Potassium Bicarbonate
Potassium Bisulfate
Potassium Bitartrate
Potassium Hydroxide
Propane
Propylene
Pyridoxine

Retinol
Riboflavin
Saccharin
Sodium Acetate
Sodium Bicarbonate
Sodium Cyclamate
Sodium Hydroxide
Sodium Polyacrylate
Styrene
Sucrose
Sucrose Polyester
Sulfuric Acid
Tannic Acid
Testosterone
Theobromine
Thiamine
Toluene
Triclocarban
Triclosan
Urea
Vanillin
Water

IODINE

Potassium Iodide
Silver Iodide

IRON

Iron(II) Oxide
Iron(III) Oxide

MAGNESIUM

Chlorophyll
Magnesium Chloride
Magnesium Hydroxide
Magnesium Oxide
Magnesium Silicate
    Hydroxide
Magnesium Sulfate

MERCURY

Mercury(II) Sulfide

NITROGEN

2,4,6-Trinitrotoluene
Acetaminophen

Ammonia
Ammonium Chloride
Ammonium Hydroxide
Ammonium Nitrate
Ammonium Sulfate
Amoxicillin
Amyl Nitrite
Caffeine
Cellulose Nitrate
Chlorophyll
Collagen
Cyanoacrylate
Cyanocobalamin
Denatonium Benzoate
Folic Acid
Gelatin
L-Aspartyl-L-Phenylalanine
  Methyl Ester
Luminol
Monosodium Glutamate
N,N-Diethyl-3-Methyl-
  benzamide
Niacin
Nicotine
Nitric Acid
Nitric Oxide
Nitrogen Dioxide
Nitroglycerin
Nylon 6 and Nylon 66
Penicillin
Polyurethane
Potassium Nitrate
Pyridoxine
Riboflavin
Saccharin
Silver Nitrate
Sodium Cyclamate
Theobromine
Thiamine
Triclocarban
Urea
Nitrous Oxide

OXYGEN

2-(4-Isobutylphenyl)propionic
  Acid
2,4,6-Trinitrotoluene
Acetaminophen

Acetic acid
Acetylsalicylic acid
Alpha-Tocopherol
Aluminum Hydroxide
Aluminum Potassium Sulfate
Aluminum Oxide
Ammonium Hydroxide
Ammonium Nitrate
Ammonium Sulfate
Amoxicillin
Amyl Acetate
Amyl Nitrite
Ascorbic Acid
Benzoic Acid
Boric Acid
Butyl Acetate
Butylated Hydroxyanisole
  and Butylated Hydro-
  xytoluene
Caffeine
Calcium Carbonate
Calcium Hydroxide
Calcium Oxide
Calcium Phosphate
Calcium Silicate
Calcium Sulfate
Camphor
Carbon Dioxide
Carbon Monoxide
Cellulose
Cellulose Nitrate
Cellulose Xanthate
Chlorophyll
Cholesterol
Cinnamaldehyde
Citric Acid
Collagen
Copper(I) Oxide
Copper(II) Oxide
Copper(II) Sulfate
Cyanoacrylate
Cyanocobalamin
Denatonium Benzoate
Dimethyl Ketone
Ethyl Acetate
Ethyl Alcohol
Ethylene Glycol
Ethylene Oxide

Folic Acid
Formaldehyde
Fructose
Gelatin
Glucose
Glycerol
Hydrogen Peroxide
Iron(II) Oxide
Iron(III) Oxide
Isoamyl Acetate
Isopropyl Alcohol
Lactic Acid
Lactose
L-Aspartyl-L-Phenylalanine
  Methyl Ester
Luminol
Magnesium Hydroxide
Magnesium Oxide
Magnesium Silicate
  Hydroxide
Magnesium Sulfate
Menthol
Methyl Alcohol
Methyl-t-butyl Ether
Monosodium Glutamate
N,N-Diethyl-3-Methyl-
  benzamide
Naproxen
Niacin
Nitric Acid
Nitric Oxide
Nitrogen Dioxide
Nitroglycerin
Nitrous Oxide
Nylon 6 and Nylon 66
Oxalic Acid
Pectin
Penicillin
Perchlorates
Petroleum
Phenol
Phosphoric Acid
Polycarbonates
Polymethyl Methacrylate
Polysiloxane
Polyurethane
Potassium Bicarbonate
Potassium Bisulfate

Potassium Bitartrate
Potassium Carbonate
Potassium Hydroxide
Potassium Nitrate
Potassium Sulfate
Pyridoxine
Retinol
Riboflavin
Saccharin
Silicon Dioxide
Silver Nitrate
Silver(I) Oxide
Sodium Acetate
Sodium Bicarbonate
Sodium Carbonate
Sodium Cyclamate
Sodium Hydroxide
Sodium Hypochlorite
Sodium Perborate
Sodium Phosphate
Sodium Polyacrylate
Sodium Silicate
Sodium Sulfite
Sodium Tetraborate
Sodium Thiosulfate
Sucrose
Sucrose Polyester
Sulfur Dioxide
Sulfuric Acid
Tannic Acid
Testosterone
Theobromine
Thiamine
Triclocarban
Triclosan
Urea
Vanillin
Water
Zinc Oxide

PHOSPHORUS

Calcium Phosphate

Phosphoric Acid
Sodium Phosphate

POTASSIUM

Aluminum Potassium
  Sulfate
Potassium Bicarbonate
Potassium Bisulfate
Potassium Bitartrate
Potassium Carbonate
Potassium Chloride
Potassium Fluoride
Potassium Hydroxide
Potassium Iodide
Potassium Nitrate
Potassium Sulfate

SILICON

Calcium Silicate
Magnesium Silicate
  Hydroxide
Polysiloxane
Silicon Dioxide
Sodium Silicate

SILVER

Silver Iodide
Silver Nitrate
Silver(I) Oxide
Silver(I) Sulfide

SODIUM

Cellulose Xanthate
Monosodium Glutamate
Sodium Acetate
Sodium Bicarbonate
Sodium Carbonate
Sodium Chloride
Sodium Fluoride
Sodium Hydroxide

Sodium Hypochlorite
Sodium Perborate
Sodium Phosphate
Sodium Polyacrylate
Sodium Silicate
Sodium Sulfite
Sodium Tetraborate
Sodium Thiosulfate

SULFUR

2,2'-Dichlorodiethyl
  Sulfide
Aluminum Potassium
  Sulfate
Ammonium Sulfate
Amoxicillin
Butyl Mercaptan
Calcium Sulfate
Cellulose Xanthate
Copper(II) Sulfate
Magnesium Sulfate
Mercury(II) Sulfide
Methyl Mercaptan
Penicillin
Potassium Bisulfate
Potassium Sulfate
Saccharin
Silver(I) Sulfide
Sodium Cyclamate
Sodium Sulfite
Sodium Thiosulfate
Sulfur Dioxide
Sulfuric Acid
Thiamine

TIN

Stannous Fluoride

ZINC

Zinc Oxide

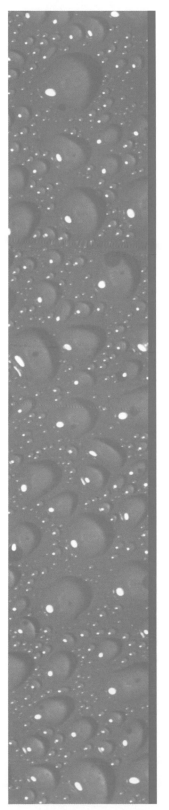

Compounds by Type

ACID

2-(4-Isobutylphenyl)propionic
  Acid
Acetic acid
Acetylsalicylic acid
Ascorbic Acid
Benzoic Acid
Boric Acid
Butyl Acetate
Citric Acid
Denatonium Benzoate
Folic Acid
Hydrogen Chloride
Lactic Acid
Naproxen
Niacin
Nitric Acid
Oxalic Acid
Penicillin
Phosphoric Acid
Potassium Bicarbonate
Potassium Bisulfate
Potassium Bitartrate
Sodium Bicarbonate
Sulfuric Acid
Tannic Acid

ALCOHOL

Ethyl Alcohol
Ethylene Glycol
Glycerol
Isopropyl Alcohol
Menthol
Methyl Alcohol
Retinol

ALDEHYDE

Cinnamaldehyde
Formaldehyde

ALKALOID

Caffeine
Nicotine
Theobromine

ALKANE

Butane
Hexane
Methane
Propane

ALKENE

1,3-Butadiene
Ethylene
Propylene

ALKYNE

Acetylene

AMIDE

Acetaminophen

BASE

Aluminum Hydroxide
Ammonia
Ammonium Hydroxide
Caffeine
Calcium Hydroxide
Magnesium Hydroxide
Potassium Hydroxide
Sodium Hydroxide
Theobromine

CARBOHYDRATE

Cellulose
Cellulose Nitrate
Fructose
Glucose
Lactose
Sucrose

CARBOXYLIC ACID

Acetic acid
Acetylsalicylic acid
Butyl Acetate
Citric Acid
Lactic Acid
Naproxen
Niacin
Oxalic Acid

ESTER

Amyl Acetate
Amyl Nitrite
Cyanoacrylate
Ethyl Acetate
Isoamyl Acetate
L-Aspartyl-L-Phenylalanine
   Methyl Ester
Nitroglycerin

ETHER

Ethylene Oxide
Methyl-t-butyl Ether
Vanillin

HYDROCARBON

1,3-Butadiene
Acetylene
Benzene
Beta-Carotene
Butane
Cumene
Ethylbenzene
Ethylene
Hexane
Isoprene
Methane
Naphthalene
Propane
Propylene
Styrene
Toluene

INORGANIC

Aluminum Fluoride
Aluminum Hydroxide
Aluminum Oxide
Aluminum Potassium Sulfate
Ammonia
Ammonium Chloride
Ammonium Hydroxide
Ammonium Nitrate
Ammonium Sulfate
Boric Acid
Calcium Carbonate
Calcium Hydroxide
Calcium Oxide
Calcium Phosphate
Calcium Silicate
Calcium Sulfate
Carbon Dioxide
Carbon Monoxide
Copper(I) Oxide
Copper(II) Oxide
Copper(II) Sulfate
Hydrogen Chloride
Iron(II) Oxide
Iron(III) Oxide
Magnesium Chloride
Magnesium Hydroxide
Magnesium Oxide
Magnesium Silicate Hydroxide
Magnesium Sulfate
Mercury(II) Sulfide
Nitric Acid
Nitric Oxide
Nitrogen Dioxide
Nitrous Oxide
Perchlorates
Phosphoric Acid
Polysiloxane
Potassium Bicarbonate
Potassium Bisulfate
Potassium Bitartrate
Potassium Carbonate
Potassium Chloride
Potassium Fluoride
Potassium Hydroxide
Potassium Iodide
Potassium Nitrate
Potassium Sulfate
Silicon Dioxide
Silver Iodide
Silver Nitrate
Silver(I) Oxide
Silver(I) Sulfide
Sodium Acetate
Sodium Bicarbonate
Sodium Carbonate
Sodium Chloride
Sodium Fluoride
Sodium Hydroxide
Sodium Hypochlorite
Sodium Perborate
Sodium Phosphate
Sodium Silicate
Sodium Sulfite
Sodium Tetraborate
Sodium Thiosulfate
Stannous Fluoride
Sulfur Dioxide
Sulfuric Acid
Water
Zinc Oxide

KETONE

Camphor
Dimethyl Ketone

METALLIC OXIDE

Aluminum Oxide
Calcium Oxide
Copper(I) Oxide
Copper(II) Oxide
Iron(II) Oxide
Iron(III) Oxide
Magnesium Oxide
Silver(I) Oxide
Zinc Oxide

NONMETALLIC OXIDE

Carbon Dioxide
Carbon Monoxide
Hydrogen Peroxide
Nitric Oxide
Nitrogen Dioxide
Nitrous Oxide
Silicon Dioxide
Sulfur Dioxide

ORGANIC

1,3-Butadiene
2-(4-Isobutylphenyl)propionic
   Acid
2,2'-Dichlorodiethyl Sulfide
2,4,6-Trinitrotoluene
Acetaminophen
Acetic acid
Acetylene
Acetylsalicylic acid
Alpha-Tocopherol
Amoxicillin
Amyl Acetate
Amyl Nitrite
Ascorbic Acid
Benzene
Benzoic Acid
Beta-Carotene
Butane
Butyl Acetate
Butyl Mercaptan
Butylated Hydroxyanisole and
   Butylated Hydroxytoluene
Caffeine
Camphor

Carbon Tetrachloride
Cellulose
Cellulose Nitrate
Cellulose Xanthate
Chloroform
Chlorophyll
Cholesterol
Cinnamaldehyde
Citric Acid
Collagen
Cumene
Cyanoacrylate
Cyanocobalamin
Denatonium
   Benzoate
Dichlorodifluoromethane
Dichlorodiphenyltrichloro-
   ethane
Dimethyl Ketone
Ethyl Acetate
Ethyl Alcohol
Ethylbenzene
Ethylene
Ethylene Glycol
Ethylene Oxide
Folic Acid
Formaldehyde
Fructose
Gamma-1,2,3,4,5,6-Hexachloro-
   cyclohexane
Glucose
Glycerol
Hexane
Hydrogen Peroxide
Isoamyl Acetate
Isoprene
Isopropyl Alcohol
Lactic Acid
Lactose
L-Aspartyl-L-Phenylalanine
   Methyl Ester
Luminol
Menthol
Methane
Methyl Alcohol
Methyl Mercaptan
Methyl-t-butyl Ether
Monosodium Glutamate

N,N-Diethyl-3-Methylbenza-
   mide
Naphthalene
Naproxen
Niacin
Nicotine
Nitroglycerin
Nylon 6 and Nylon 66
Oxalic Acid
Penicillin
Phenol
Poly(Styrene-Butadiene-
   Styrene)
Polycarbonates
Polyethylene
Polymethyl Methacrylate
Polypropylene
Polystyrene
Polytetrafluoroethylene
Polyurethane
Polyvinyl Chloride
Propane
Propylene
Pyridoxine
Retinol
Riboflavin
Saccharin
Sodium Cyclamate
Sodium Polyacrylate
Styrene
Sucrose
Sucrose Polyester
Tannic Acid
Testosterone
Theobromine
Thiamine
Toluene
Triclocarban
Triclosan
Urea
Vanillin

PHENOL

Butylated Hydroxyanisole
   and Butylated Hydroxy-
   toluene
Phenol

POLYMER

Cellulose
Cellulose Nitrate
Cellulose Xanthate
Nylon 6 and Nylon 66
Poly(Styrene-Butadiene-
    Styrene)
Polycarbonates
Polyethylene
Polymethyl Methacrylate
Polypropylene
Polysiloxane
Polystyrene
Polytetrafluoroethylene
Polyurethane
Polyvinyl Chloride
Sodium Polyacrylate

SALT

Aluminum Fluoride
Aluminum Potassium Sulfate
Ammonium Chloride
Ammonium Nitrate
Ammonium Sulfate
Calcium Carbonate
Calcium Phosphate
Calcium Silicate
Calcium Sulfate
Copper(II) Sulfate
Magnesium Chloride
Magnesium Silicate
    Hydroxide
Magnesium Sulfate
Mercury(II) Sulfide
Monosodium Glutamate
Perchlorates
Potassium Bicarbonate
Potassium Bisulfate
Potassium Bitartrate
Potassium Carbonate
Potassium Chloride
Potassium Fluoride
Potassium Iodide
Potassium Nitrate
Potassium Sulfate
Silver Iodide
Silver Nitrate
Silver(I) Sulfide
Sodium Acetate
Sodium Bicarbonate
Sodium Carbonate
Sodium Chloride
Sodium Cyclamate
Sodium Fluoride
Sodium Hypochlorite
Sodium Perborate
Sodium Phosphate
Sodium Silicate
Sodium Sulfite
Sodium Tetraborate
Sodium Thiosulfate
Stannous Fluoride

VITAMIN

Alpha-Tocopherol
Ascorbic Acid
Cyanocobalamin
Folic Acid
Niacin
Pyridoxine
Retinol
Riboflavin
Thiamine

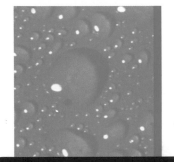

for further information

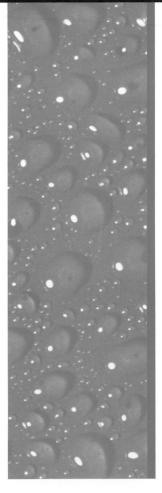

BOOKS

Attenborough, David. *The Private Life of Plants.* Princeton, NJ: Princeton University Press, 1995.

Brody, Tom. *Nutritional Biochemistry.* San Diego: Academic Press, 1998.

Buchanan, B. B., W. Gruissem, and R. L. Jones. *Biochemistry and Molecular Biology of Plants.* Rockville, MD: American Society of Plant Physiologists, 2000.

Buechel, K. H., et al. *Industrial Inorganic Chemistry.* New York: VCH, 2000.

Buschmann, Helmut, et al. *Analgesics: From Chemistry and Pharmacology to Clinical Application.* New York: Wiley-VCH, 2002.

Butler, A. R., and R. Nicholson. *Life, Death and Nitric Oxide.* London: Royal Society of Chemistry, 2003.

Cagin, Seth, and Philip Dray. *Between Earth and Sky: How CFCs Changed Our World and Endangered the Ozone Layer.* New York: Pantheon Books, 1993.

Carpenter, Kenneth J. *The History of Scurvy and Vitamin C.* Cambridge, UK: Cambridge University Press, 1986.

Carson, Rachel. *Silent Spring.* Boston: Houghton Mifflin, 1962.

Cavitch, Susan Miller. *The Natural Soap Book: Making Herbal and Vegetable-Based Soaps.* Markham, Canada: Storey Publishing, 1995.

Challem, Jack, and Melissa Diane Smith. *Basic Health Publications User's Guide to Vitamin E: Don't Be a Dummy: Become an Expert on What Vitamin E Can Do for Your Health.* North Bergen, NJ: Basic Health Publications, 2002.

Chalmers, Louis. *Household and Industrial Chemical Specialties.* Vol. 1. New York: Chemical Publishing Co., Inc., 1978.

Cherniske, Stephen. *Caffeine Blues: Wake Up to the Hidden Dangers of America's #1 Drug.* New York: Warner Books, 1998.

"Cholesterol, Other Lipids, and Lipoproteins." In *In-Depth Report.* Edited by Julia Goldrosen. Atlanta: A.D.A.M., 2004.

Cooper, P. W., and S. R. Kurowski. *Introduction to the Technology of Explosives.* New York: Wiley-VCH, 1997.

Cornell, Rochelle M., and Udo Schwertmann. *The Iron Oxides: Structure, Properties, Reactions, and Uses*, 2nd ed. New York: Wiley-VCH, 2003.

CRC Handbook of Chemistry and Physics. David R. Lide, editor in chief. 86th ed. Boca Raton, FL: Taylor & Francis, 2005.

Dean, Carolyn. *The Miracle of Magnesium.* New York: Ballantine Books, 2003.

Dick, John S., and R. A. Annicelli, eds. *Rubber Technology: Compounding and Testing for Performance.* Cincinnati, OH: Hanser Gardner Publications, 2001.

Dunlap, Thomas. *DDT: Scientists, Citizens, and Public Policy.* Princeton, NJ: Princeton University Press, 1983.

Dwyer, Bob, et al. *Carbon Monoxide: A Clear and Present Danger.* Mount Prospect, IL: ESCO Press, 2004.

Eades, Mary Dan. *The Doctor's Complete Guide to Vitamins and Minerals.* New York: Dell, 2000.

Environment Canada, Health Canada. *Ethylene Oxide.* Ottawa: Environment Canada, 2001.

Food Antioxidants: Technological, Toxicological, and Health Perspectives, D. L. Madhavi, S. S. Deshpande, and D. K. Salunkhe, eds. New York: Dekker, 1996.

Gahlinger, Paul M. *Illegal Drugs: A Complete Guide to Their History.* Salt Lake City, UT: Sagebrush Press, 2001.

Genge, Ngaire E. *The Forensic Casebook: The Science of Crime Scene Investigation.* New York: Ballantine, 2002.

Grimm, Tom, and Michele Grimm. *The Basic Book of Photography.* New York: Plume Books, 2003.

Harte, John, et al. *Toxics A to Z.* Berkeley: University of California Press, 1991.

Hermes, Matthew E. *Enough for One Lifetime: Wallace Carothers, Inventor of Nylon.* Philadelphia: Chemical Heritage Foundation, 1996.

Holgate, S. T., et al. *Air Pollution and Health.* New York: Academic Press, 1999.

Hyne, Norman J. *Nontechnical Guide to Petroleum Geology, Exploration, Drilling and Production*, 2nd ed. Tulsa, OK: Pennwell Books, 2001.

Jeffreys, Diarmuid. *Aspirin: The Remarkable Story of a Wonder Drug.* New York: Bloomsbury, 2004.

Johnson, Peter S. *Rubber Processing: An Introduction.* Cincinnati, OH: Hanser Gardner Publications, 2001.

Karger-Kocsis, J. *Polypropylene—An A-Z Reference.* New York: Springer-Verlag, 1998.

Kirk-Othmer Encyclopedia of Chemical Technology, 4th ed. New York: John Wiley & Sons, 1991.

Knox, J. Paul, and Graham B. Seymour, eds. *Pectins and Their Manipulation.* Boca Raton, FL: CFC Press, June 2002.

Kurlansky, Mark. *Salt: A World History.* New York: Walker, 2002.

Leffler, William L. *Petroleum Refining in Nontechnical Language.* Tulsa, OK: Pennwell Books, 2000.

Mead, Clifford, and Thomas Hager, eds. *Linus Pauling: Scientist and Peacemaker.* Portland, OR: Oregon State University Press, 2001.

Mebane, Robert C., and Thomas R. Rybolt. *Plastics and Polymers.* New York: Twenty-First Century, 1995.

Meikle, Jeffrey L. *American Plastic: A Cultural History.* New Brunswick, NJ: Rutgers University Press, 1995.

Menhard, Francha Roffe. *The Facts about Inhalants.* New York: Benchmark Books, 2004.

Misra, Chanakya. *Industrial Alumina Chemicals.* Washington, DC: American Chemical Society, 1986.

Mosby's Medical, Nursing, and Allied Health Dictionary, 5th ed. St. Louis: Mosby, 1998.

Multhauf, Robert P., and Christine M. Roane. "Nitrates." In *Dictionary of American History*. 3rd ed., vol. 6. Stanley I. Kutler, ed. New York: Charles Scribner's Sons, 2003.

Nabors, Lyn O'Brien, ed. *Alternative Sweeteners (Food Science and Technology)*, 3rd rev. London: Marcel Dekker, 2001.

Packer, Lester, and Carol Colman. *The Antioxidant Miracle: Put Lipoic Acid, Pycogenol, and Vitamins E and C to Work for You.* New York: Wiley, 1999.

Patnaik, Pradyot. *Handbook of Inorganic Chemicals.* New York: McGraw-Hill, 2003.

Rain, Patricia. *Vanilla: A Cultural History of the World's Most Popular Flavor and Fragrance.* Edited by Jeremy P. Tarcher. New York: Penguin Group USA, 2004.

Rainsford, K. D., ed. *Ibuprofen: A Critical Bibliographic Review.* Bethesda, MD: CCR Press, 1999.

Richardson, H. W., ed. *Handbook of Copper Compounds and Applications.* New York: Marcel Dekker, 1997.

Sherman, Josepha, and Steve Brick. *Fossil Fuel Power.* Mankato, MN: Capstone Press, 2003.

Snyder, C. H. *The Extraordinary Chemistry of Ordinary Things,* 4th ed. New York: John Wiley and Sons, 2002.

Strange, Veronica. *The Meaning of Water.* Oxford, UK: Berg Publishers, 2004.

Stratmann, Linda. *Chloroform: The Quest for Oblivion.* Phoenix Mill, UK: Sutton Publishing Co., 2003.

Tegethoff, F. Wolfgana, with Johannes Rohleder and Evelyn Kroker, eds. *Calcium Carbonate: From the Cretaceous Period into the 21st Century.* Boston: Birkhäuser Verlag, 2001.

Tyman, J. H. P. *Synthetic and Natural Phenols.* Amsterdam: Elsevier, 1996.

Ware, George W. *The Pesticide Book.* Batavia, IL: Mesiter, 1999.

Weinberg, Alan Bennet, and Bonnie K. Bealer. *The World of Caffeine: The Science and Culture of the World's Most Popular Drug.* New York: Routledge, 2002.

Weissermel, Klaus, and Hans-Jürgen Arpe. *Industrial Organic Chemistry.* Weinheim, Germany: Wiley-VCH, 2003, 117-122.

Wyman, Carolyn. *JELL-O: A Biography.* Fort Washington, PA: Harvest Books, 2001.

PERIODICALS

"Another Old-Fashioned Product Vindicates Itself." *Medical Update* (October 1992): 6.

Arnst, Catherine. "A Preemptive Strike against Cancer." *Business Week* (June 7, 2004): 48.

Baker, Linda. "The Hole in the Sky (Ozone Layer)." *E* (November 2000): 34.

Bauman, Richard. "Getting Skunked: Understanding the Antics behind the Smell." *Backpacker* (May 1993): 30-31.

Drake, Geoff. "The Lactate Shuttle—Contrary to What You've Heard, Lactic Acid Is Your Friend." *Bicycling* (August 1992): 36.

Fox, Berry. "Not Fade Away." *New Scientist* (March 1, 2003); 40.

"Global Ethyl, Butyl Acetate Demand Expected to Rebound." *The Oil and Gas Journal* (April 24, 2000): 27.

Gorman, Christine. "The Bomb Lurking in the Garden Shed." *Time* (May 1, 1995): 54.

Karaim, Reed. "Not So Fast with the DDT: Rachel Carson's Warnings Still Apply." *American Scholar* (June 2005): 53-60.

Keenan, Faith. "Blocking Liver Damage." *Business Week* (October 21, 2002): 147.

Kluger, Jeffrey. "The Buzz on Caffeine." *Time* (December 20, 2004): 52.

Lazear, N. R. "Polycarbonate: High-Performance Resin." *Advanced Materials & Processes* (February 1995): 43-45.

Legwold, Gary. "Hydration Breakthrough." *Bicycling* (July 1994): 72-73.

Liu, Guanghua. "Chinese Cinnabar." *The Mineralogical Record* (January-February 2005): 69-80.

Malakoff, David. "Public Health: Aluminum Is Put on Trial as a Vaccine Booster." *Science* (May 26, 2000): 1323.

Mardis, Anne L. "Current Knowledge of the Health Effects of Sugar Intake." *Family Economics and Nutrition Review* (Winter 2001): 88-91.

McGinn, Anne Platt. "Malaria, Mosquitoes, and DDT: The Toxic War against a Global Disease." *World Watch* (May 1, 2002): 10-17.

Metcalfe, Ed, et al. "Sweet Talking." *The Ecologist* (June 2000): 16.

Milius, Susan. "Termites Use Mothballs in Their Nests." *Science News* (May 2, 1998): 228.

Neff, Natalie. "No Laughing Matter." *Auto Week* (May 19, 2003): 30.

"Nitroglycerin: Dynamite for the Heart." *Chemistry Review* (November 1999): 28.

O'Neil, John. "And It Doesn't Taste Bad, Either." *New York Times* (November 30, 2004): F9.

Pae, Peter. "Sobering Side of Laughing Gas." *Washington Post* (September 16, 1994): B1.

Palvetz, Barry A. "A Bowl of Hope, Bucket of Hype?" *The Scientist* (April 2, 2001): 15.

Rawls, Rebecca. "Nitroglycerin Explained." *Chemical & Engineering News* (June 10, 2002): 12.

Rowley, Brian. "Fizzle or Sizzle? Potassium Bicarbonate Could Help Spare Muscle and Bone." *Muscle & Fitness* (December 2002): 72.

Russell, Justin. "Fuel of the Forgotten Deaths." *New Scientist* (February 6, 1993): 21-23.

Schramm, Daniel. "The North American USP Petrolatum Industry." *Soap & Cosmetics*, (January 2002): 60-63.

Stanley, Peter. "Nitric Oxide." *Biological Sciences Review* (April 2002): 18-20.

"The Stink that Stays." *Popular Mechanics* (December 2004): 26.

Strobel, Warren P. "Saddam's Lingering Atrocity." *U.S. News & World Report* (November 27, 2000): 52.

"Strong Muscle and Bones." *Prevention* (June 1, 1995): 70-73.

"Taking Supplements of the Antioxidant." *Consumer Reports* (September 2003): 49.

Travis, J. "Cool Discovery: Menthol Triggers Cold-Sensing Protein." *Science News* (February 16, 2002): 101-102.

U.S. Department of Health and Human Services. "Methanol Toxicity." *American Family Physician* (January 1993): 163-171.

"Unusual Thermal Defence by a Honeybee against Mass Attack by Hornets." *Nature* (September 28, 1995): 334-336.

"USDA Approves Phosphate to Reduce Salmonella in Chicken." *Environmental Nutrition* (February 1993): 3.

Vartan, Starre. "Pretty in Plastic: Pleather Is a Versatile, though Controversial, Alternative to eather." *E* (September-October 2002): 53-54.

"Vitamins: The Quest for Just the Right Amount." *Harvard Health Letter* (June 2004): 1.

Walter, Patricia A. "Dental Hypersensitivity: A Review." *The Journal of Contemporary Dental Practice* (May 15, 2005): 107-117.

Young, Jay A. "Copper (II) Sulfate Pentahydrate." *Journal of Chemical Education* (February 2002): 158.

WEBSITES

"The A to Z of Materials." Sydney, Australia: Azom.com. http://www.azom.com/ (accessed on March 1, 2006).

"Air Toxics Website." U.S. Environmental Protection Agency Technology Transfer Network http://www.epa.gov/ttn/atw/ (accessed on March 10, 2006).

Calorie Control Council. Atlanta, GA: Calorie Control Council. http://www.caloriecontrol.org/

CHEC's Health*e*House: The Resource for Environmental Health Risks Affecting Your Children. Los Angeles, CA: Children's Health Environmental Coalition. http://www.checnet.org/ehouse (accessed on March 13, 2006).

Chemfinder.com. Cambridge, MA: CambridgeSoft Corporation. http://www.chemfinder.cambridgesoft.com (accessed on March 13, 2006).

"Chemical Backgrounders." Itasca, IL: National Safety Council. http://www.nsc.org/library/chemical/ (accessed on March 1, 2006).

"Chemical Profiles." Scorecard. The Pollution Information Site. Washington, DC: Green Media Toolshed. http://www.scorecard.org/chemical-profiles/ (accessed on March 1, 2006).

The Chemistry Store. Cayce, SC: ChemistryStore.com, Inc. http://www.chemistrystore.com/ (accessed on March 13, 2006).

Chemistry.org. Washington, DC: American Chemical Society. http://www.chemistry.org/portal/a/c/s/1/home.html (accessed on March 13, 2006).

Cheresources.com: Online Chemical Engineering Information. Midlothian, VA: The Chemical Engineers' Resource Page. http://www.cheresources.com (accessed on March 13, 2006).

"Dietary Supplement Fact Sheets." Bethesda, MD: U.S. National Institutes of Health, Office of Dietary Supplements. http://ods.od.nih.gov/Health_Information/Information_About_Individual_Dietary_Supplements.aspx (accessed on March 13, 2006).

Drugs.com. Auckland, New Zealand: Drugsite Trust. http://www.drugs.com (accessed on March 13, 2006).

EnvironmentalChemistry.com. Portland, ME: Environmental Chemistry.com. http://environmentalchemistry.com (accessed on March 13, 2006).

Exploratorium: the museum of science, art and human perception. San Francisco, CA: Exploratorium at the Palace of Fine Arts. http://www.exploratorium.edu/ (accessed on March 13, 2006).

ExToxNet: The Extension Toxicology Network. Corvallis, OR: Oregon State University. http://extoxnet.orst.edu/ (accessed on March 13, 2006).

Fibersource: The Manufactured Fibers Industry. Arlington, VA: Fiber Economics Bureau. http://www.fibersource.com (accessed on March 13, 2006).

General Chemistry Online. Frostburg, MD: Frostburg State University, Department of Chemistry. http://antoine.frostburg.edu/chem/senese/101/index.shtml (accessed on March 13, 2006).

Household Products Database. Bethesda, MD: U.S. National Library of Medicine. http://householdproducts.nlm.nih.gov/ (accessed on March 13, 2006).

Integrated Risk Information System. Washington, DC: U.S. Environmental Protection Agency. http://www.epa.gov/iris/index.html (accessed on March 13, 2006).

"International Chemical Safety Cards (ICSCs)." Geneva, Switzerland: International Labour Organization. http://www.ilo.org/public/english/protection/safework/cis/products/icsc/index.htm (accessed on March 1, 2006).

"International Chemical Safety Cards." Atlanta, GA: U.S. National Institute for Occupational Safety and Health. http://www.cdc.gov/niosh/ipcs/icstart.html (accessed on March 1, 2006).

IPCS INTOX Databank. Geneva, Switzerland: International Programme on Chemical Safety. http://www.intox.org/ (accessed on March 13, 2006).

Kimball's Biology Pages. Andover, MA: John W. Kimball. http://biology-pages.info (accessed on March 13, 2006).

Linus Pauling Institute Micronutrient Information Center. Corvallis, OR: Oregon State University, Linus Pauling Institute. http://lpi.oregonstate.edu/infocenter/ (accessed on March 13, 2006).

MadSci Network. Boston, MA: Third Sector New England. http://www.madsci.org (accessed on March 13, 2006).

Medline Plus. Bethesda, MD: U.S. National Library of Medicine. http://www.nlm.nih.gov/medlineplus (accessed on March 1, 2006).

Mineral Information Institute. Golden, CO: Mineral Information Institute. http://www.mii.org (accessed March 13, 2006).

Mineralogy Database. Spring, TX: Webmineral.com. http://webmineral.com/ (accessed on March 13, 2006).

"Molecule of the Month Page." Bristol, United Kingdom: University of Bristol School of Chemistry. http://www.chm.bris.ac.uk/motm/motm.htm (accessed on March 1, 2006).

National Historic Chemical Landmarks. Washington, DC: American Chemical Society. http://acswebcontent.acs.org/landmarks/index.html (accessed on March 13, 2006).

NIST Chemistry Webbook. Gaithersburg, MD: U.S. National Institute of Standards and Technology. http://webbook.nist.gov (accessed on March 13, 2006).

PAN Pesticides Database. San Francisco, CA: Pesticide Action Network, North America. http://www.pesticideinfo.org (accessed March 13, 2006).

"Polymer Science Learning Center." Hattiesburg, Mississippi: University of Southern Mississippi, Department of Polymer Science. http://www.pslc.ws/ (accessed on March 13, 2006).

PubChem. Bethesda, MD: U.S. National Library of Medicine. http://pubchem.ncbi.nlm.nih.gov/ (accessed on March 13, 2006).

Reciprocal.net. Bloomington, IN: Indiana University Molecular Structure Center. http://www.reciprocalnet.org/ (accessed on March 13, 2006).

The Science Center: A Teacher's Guide to Educational Resources on the Internet. Arlington, VA: Chlorine Chemistry Council. http://www.scienceeducation.org (accessed on March 13, 2006).

Shakhashiri, Bassam Z. "Chemical of the Week." Science Is Fun. Madison, WI: University of Wisconsin. http://scifun.chem.wisc.edu/chemweek/chemweek.html (accessed on March 13, 2006).

3Dchem.com. Oxford, United Kingdom: University of Oxford, Department of Chemistry. http://www.3dchem.com (accessed on March 13, 2006).

"ToxFAQs^{TM}" Atlanta, GA: U.S. Agency for Toxic Substances and Disease Registry. http://www.atsdr.cdc.gov/toxfaq.html (accessed on March 13, 2006).

"Toxicological Profile Information Sheets." Atlanta, GA: U.S. Agency for Toxic Substances and Disease Registry. http://www.atsdr.cdc.gov/toxpro2.html (accessed on March 1, 2006).

TOXNET: Toxicology Data Network. Bethesda, MD: U.S. National Library of Medicine. http://toxnet.nlm.nih.gov (accessed on March 13, 2006).

World of Molecules Home Page. http://www.worldofmolecules.com/ (accessed on March 13, 2006).

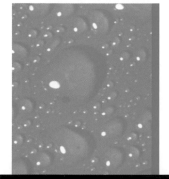

index

This index is sorted
word-by-word. Italic
type indicates volume
numbers; **boldface**
indicates main
entries; (ill.) indicates
illustrations.

D

G

Vaccines, *1*:47
Vanillin, *3*:**873-877**, 874 (ill.)
Vaseline, *2*:547-549
Vasodilators, *1*:89-90
Vermillion, *2*:440
Vicks VapoRub, *1*:173; *2*:437-438
Vinegar, *1*:23-26
Viscose Corp. of America, *1*:208
Vision, *3*:679
Vitamins
    alpha-tocopherol (E), *1*:37-40
    ascorbic acid (C), *1*:93-97
    beta-carotene, *1*:109-113; *3*:678, *3*:680
    cyanocobalamin (B_{12}), *1*:265-268
    folic acid (B_9), *2*:321-324
    niacin (B_3), *2*:483-486
    pyridoxine (B_6), *3*:673-676
    retinol (A), *3*:677-681
    riboflavin (B_2), *3*:683-687
    thiamine (B_1), *3*:822, 847-851
    See also Minerals (nutritional)
Volpenhein, Robert, *3*:813
Vonnegut, Bernard, *3*:703

Wackenroder, Heinrich Wilhelm
    Ferdinand, *1*:110
Waldie, David, *1*:211
Warning odors, *1*:87
Washing soda, *3*:729-733
Water, *3*:**879-884**, 880 (ill.)
Water treatment
    aluminum potassium sulfate, *1*:55

ammonium sulfate, *1*:78
    fluoride, *1*:43; *3*:748-749
    sodium sulfite in, *3*:786
    softeners, *3*:737
Wedgwood, Thomas, *3*:706
Wehmer, Carl, *1*:235
Wells, Horace, *2*:514
Whipple, George Hoyt, *1*:266
Whitehall Laboratories, *1*:11
Wiegleb, Johann Christian, *2*:525
Wieland, Heinrich Otto, *1*:224-225
Wilbrand, Joseph, *1*:15
Williams, Robert R., *3*:849-850
Willow bark, *1*:31-32
Wills, Lucy, *2*:321
Winhaus, Adolf, *1*:224-225
Wislicenus, Johannes, *2*:392
Wöhler, Friedrich, *1*:27; *2*:525; *3*:867
Wood alcohol, *2*:449-453
Wood preservatives, *1*:248, 252
Woodward, Robert Burns, *1*:219, 225, 267
Wounds, sealing of, *1*:260-261
Wulff process, *1*:28
Wurtz, Charles Adolphe, *2*:313, 317

Xylene, *3*:855

Zeidler, Othmar, *1*:283
Zeolites, *1*:257
Ziegler, Karl, *2*:384, 580, *2*:582, 588-589
Zinc oxide, *3*:**885-888**, 886 (ill.)